Plastics, Rubber and Health

Guneri Akovali

Smithers Rapra Technology Limited
A wholly owned subsidiary of The Smithers Group

Shawbury, Shrewsbury, Shropshire, SY4 4NR, United Kingdom
Telephone: +44 (0)1939 250383 Fax: +44 (0)1939 251118
http://www.rapra.net

First Published in 2007 by

Smithers Rapra Technology Limited

Shawbury, Shrewsbury, Shropshire, SY4 4NR, UK

Soft-backed ISBN: 978-1-84735-081-7
Hard-backed ISBN: 978-1-84735-082-4

Typeset by Smithers Rapra Technology Limited
Printed and bound by Lightning Source

Contents

This book is dedicated to

The Middle East Technical University (METU) of Ankara, Turkey

For their 50th Anniversary, (1957-2007)

Preface

Plastics and rubbers are two very different, important materials in this modern age that are needed and used a great deal in our everyday life, both in the indoors and the outdoors of our living spaces. However, there is still, controversy almost all the time, about the use of certain ones and hence there is somewhat of a misconception surrounding the use of these materials, with certain beliefs and scare stories about their possible negative effects on human health; i.e., the *belief that any form of PVC use can cause cancer*. However, today, it is known that it is not PVC itself, but its monomer, vinyl chloride, and certain special additives incorporated with it (i.e., certain phthalates used as plasticisers), which are responsible for these negative effects. And the reality is such that, PVC is being used as blood bags in medical applications successfully, with no problems. Hence, the separation line of 'whether something is toxic and harmful to health' or 'not', (and if it is, under which conditions) is a very critical issue and therefore, there needs to be a better understanding of these systems.

This book aims to present the available information on the 'plastics and rubbers and health triangle', to help to obtain a better understanding of the facts.

After some basic concepts and definitions (Chapter 2), a general toxicity approach for plastics and rubbers is presented in Chapter 3. This discusses the general toxicity issue for plastics and rubbers due to additives (in the same chapter), followed by a consideration of health effects due to the polymers and other main ingredients (except additives), in the following chapter.

After this general information, possible health effects of plastics and rubbers use in contact with food, are briefly discussed in Chapter 5.

Possible health effects from other applications of these systems are also briefly discussed in the following chapters (their use in healthcare, construction, sports and leisure, automotive and transportation, agriculture, electrical and electronic applications, and consumer products), in Chapters 6, 7 and 8.

Chapter 9 deals with sustainability through plastics and rubbers, while a comprehensive list of some health hazard causing solvents, monomers and chemicals common for plastics and rubbers are presented in Chapter 10. Chapter 11, gives a brief account of some extremely hazardous substances and carcinogens related to plastics and rubbers.

Certainly the book cannot be considered to be complete, if one considers the topic and the dynamic characteristics involved. Additional up-to-date information, whenever needed, can be obtained through the web addresses as well as the references provided at the end of each chapter.

I specifically want to thank to Frances Gardiner, the Commissioning Editor of Smithers Rapra Technology Limited, UK, for her valuable encouraging efforts and support, and for the cooperation given at all times, as well as for finalising the book in a rather short period of time.

I must also acknowledge the kind help of Vicky Tweddle, the Assistant Editor, and Sandra Hall who typeset the book and Steve Barnfield who designed the cover so nicely.

Professor Guneri Akovalı
Ankara, October 29 2007

1 Introduction

1.1 Introduction

Today there are more than 70,000 synthetic chemicals in existance, with an estimated 1,000 new chemicals being introduced every year. Most of these chemicals are labelled as 'hazardous to human health' or 'toxic', although, unfortunately only about 7% of them have ever been tested adequately to determine their effect on humans and on other forms of life.

Both indoors, where we live and spend most of our time, and outdoors, and through the food that we eat, we come into contact with, we use, we inhale and we ingest a number of these chemicals directly or indirectly, any of which can affect human health in different ways and to different extents.

These effects can be direct or indirect, and although we humans are in fact aware of them and the danger involved, the facts are surprisingly still ignored in some cases. It is interesting to cite one recent example: on 11 January 2006, ministers of environment and business and some political leaders from six leading Asia-Pacific countries (USA, China, India, Japan, South Korea, and Australia, which account for almost half of the world's energy use, with Australia and the USA, on a *per capita* basis, being the two most industrialised nations with the highest emissions) met in Sydney for a two-day conference to discuss alternative industry-focused strategies to reduce or eliminate toxic substances and pollution in the environment. This was complementary to the Kyoto Protocol, which mandates emission cuts for industrialised nations, is ratified by 145 nations, and came into effect in February 2006. It is really very interesting to note that Australia and the USA are not yet parties to the Kyoto Protocol, Japan is a supporter with binding commitments, and China (with the fastest-growing of the world's top six economies), India and South Korea are also Kyoto supporters, with non-binding commitments! Although the USA is not a participant of Kyoto, US Energy Secretary, Samuel Boldman said '...*the people..., who run these [chemical] companies, they do [also] have children, they do have grandchildren, they [all] do live and breathe in the [same] world*' [1]. One can extend these words to include all toxic substances existing around us, even in our food, to say freely: 'beware of the toxic substances all around us, the toxic substances that endanger life on earth, specifically human health, by realising that no human has immunity to them.'

An ever-increasing number of endangered species are close to extinction, in addition to those that are already extinct, due to pollution, toxic substances, and environmental conditions. Nobody is sure when humans will be added to this endangered-species list although there is the spectacular ecological (but obviously still inefficient) success achieved in general, globally during the last 30 years.

On the other hand, there is the over-increase of world population and its related problems: it has been claimed that humans have already exceeded the planet's ability to sustain their level of consumption, known as 'Earth's carrying capacity', by about 20%. That figure is expected to climb as more than 2 billion people living in India, China and other developing countries raise their standards of living in the near future, which will certainly cause ever-increasing pollution and toxic effects, unless more careful measures are planned and put in action.

Amongst the causes and consequences of environmental pollution, the 'green house effect' is identified most often. The change in global climate and the danger of turning to an Arctic age (irreversibly) triggered by global warming (irreversibly) predicted as a result, is considered one of biggest problems facing the world: 'Before this century is over, billions of us (humans) will die, and the few breeding pairs of people that survive will be in the Arctic where the climate remains tolerable ...' [2].

However, there is another and mostly neglected issue of *environmental pollutants* and *all other toxic substances* around us causing alarmingly detrimental effects on human health within much shorter periods of time, even within a human lifetime. Some of these effects are (most probably) due to the materials we use in our daily life, and plastics and rubbers constitute many of them. From plastic window profiles and doors made of polyvinyl chloride (PVC) to plastic packaging films of polyethylene, our life is very much entangled with plastics. *But there is also the common belief that any form of PVC use can cause cancer.* In spite of general ignorance about their possible effects on health, there has been controversy and sometimes misconception over these materials. It is known today that it is not the basic polymers themselves that are responsible for most of these health effects, but rather all of those foreign materials and additives that are added to them for different purposes during their processing and afterwards, which together constitute plastics and rubber systems; nor are plastics and rubbers a unique case as the same result applies to many materials that are treated for use, e.g., 'paper' is essentially an inert material, but its coated version (used to give grease resistance to microwave popcorn bags, fast food and candy wrappers, and pizza box liners), can leach out certain toxic chemicals into the food in contact with it, and these can break down to chemicals that are carcinogenic in the body [3]. A number of such examples can be cited. Hence by considering plastics and rubbers with regard to their possible effects on health, similar problems of using any other impure material will also be addressed to some extent.

Since the need for plastics and rubber materials is widespread, for a number of reasons, and since we are certainly using them and will continue to use them into the future in ever greater quantities including those materials whose use is criticised, it is very important to learn the facts and to know to what extent these criticisms are well-founded. To this end, this book will present a classification, with some details for most of the additives that are used as well as the possible evolution of new (toxic) interactions. By avoiding slogans of any sort and by trying to be aware of arguments both for and against *the aim will be just to shed some further light of scientific fact to understand the issue better.*

Except for certain specific examples given, it is expected that the reader will, in general, draw his own conclusions about any possible health issues from the evidence presented.

To begin with, this book gives information about 'chemicals that can cause health hazards' and 'toxic compounds', with regard to plastics and rubbers (mainly in Chapters 2, 3 and 10) and also carcinogenic chemicals, for their relevance to the topic. Some specific examples are described in more detail in the other chapters.

This book cannot, in one volume offer comprehensive coverage of all related subjects. However, for those seek more information, up-to-date references are provided after each chapter. The appendices in Chapters 10 and 11 will also be helpful for this purpose.

In some parts of the book certain subject areas might be addressed more than once, and due to their importance these are kept as they are, which it is hoped that the reader may tolerate; and although it cannot be complete by any means, still the book, in its existing format, should help greatly in understanding and in filling the big gap between reality and criticism.

Since we have no better material to use in place of plastics and rubbers, we must find ways to live with them, keeping their benefits while avoiding the problems involved, as far as possible. Nevertheless, it is hoped that the book will at least be successful in inspiring readers and raising awareness of the subject.

References

1.   AP Report, *Two Day Energy Conference*, Sydney, Australia, 11 January 2006.

2.   J. Lovelock, *The Independent*, 16 January 2006.

3.   E. Weise, *USA Today*, 8A, 18 November 2005.

2 Some Basic Concepts and Definitions

2.1 Plastics, Rubbers and Health

Some newly processed plastics and rubbers smell distinctly offensive. A number of volatile organic chemicals (VOC) in plastics and rubbers can be responsible for these strong chemical smells, including various additives used during their processing for different purposes, such as: amines, phenols, mercaptans, peroxides, aldehydes, ketones, alcohols, etc., in addition to some plasticisers, stabilisers and retardants, and some of the organic solvents, as well as remnants (of monomers like styrene and acrylic esters, catalysts used, etc., all of which may be left in the system after polymerisation/vulcanisation, in trace amounts), and new interaction chemicals that can be produced during processing, etc., any of these can usually be discovered because they give off a strong chemical smell, and are usually toxic (there are a number of measures that can be taken to eliminate or reduce these unpleasant odours, such as replacing them with odour-free or odour-reduced substitutes, minimising the remnants of monomers, adding proper odour absorbers or antimicrobial agents, and even using fragrances to mask them [1, 2]).

Fogging of the inside of car windows is a common problem, especially when the car is new, which is also due to the VOC emitted from either the polyurethane (PU) foam in the seats, or from the polyvinyl chloride (PVC) seat covers, etc. A new rubber tyre usually smells awful. The unpleasant smells and fogging, because they signify the emission of various VOC from these processed items, are most probably also signs of toxicants, and they can be identified and measured by standard *instrumental analytical* methods (such as gas chromatography/mass spectroscopy (GC/MS) or gas chromatography/differential mobility spectrometry (GC/DMS)), and better by *electronic noses* that rely on electronic gas sensor arrays and pattern recognition technology, at very low concentration levels of ppm (parts per million) or even ppb (parts per billion), which are well documented [3, 4]. Electronic noses (e-noses), for example, are specifically developed for robots, and can detect and locate the source of these emissions.

Some additional information for VOC are provided in Appendix 2.A.1 (at the end of this chapter).

It should be noted that, in addition to the possible emission of volatile toxic compounds from plastics and rubber materials, there is also the issue of migration of these compounds (into food) in the field of food packaging, which, at least, can impact the visual impression of packaged products as well as their organoleptic qualities [4].

The health hazards of VOC emissions, as well as the possible migration of various toxic chemicals from plastics and rubber materials, are also discussed in some more depth in Chapter 3, and in Chapters 4, 6 and 7.

2.2 A Brief Account of 'Chemicals' and 'Human Health'

Our environment contains various toxic chemicals. Beginning indoors, where we live and spend most of our time, and in the materials we use and come into contact with, there are a number of chemicals, any of which can affect human health in different ways and to different extents.

Humans, and most vertebrate animals, may be seriously affected by exposure to certain chemicals under certain conditions, directly or indirectly; these may constitute serious health hazards (leading to diseases), as well as physical hazards, including risk of death.

The US Federal Organization of Safety and Health Administration (OSHA) categorises these chemicals in general, as follows:

a) **Chemicals hazardous to health** are mainly 'carcinogens, toxic/highly toxic agents, reproductive toxics, irritants, corrosives, etc., that can act on certain organs or systems in the human body', so that they may 'damage' lungs, skin, eyes, or mucous membranes,

b) **Physically hazardous chemicals** are 'combustible liquids', 'compressed gases', 'an explosive', 'a flammable chemical' etc., depending on whether the material poses any risk of fire or explosion.

Any of these substances can be a solid, a liquid, a gas, a vapour (which is the gaseous phase of a liquid or solid material at ambient temperature and pressure), an aerosol (a dispersion of microscopic solid or liquid particles in a gaseous medium), a smoke (an aerosol usually of carbon particles with less than 0.1 μm diameter), or a fog (a visible liquid aerosol).

In this book, *hazards from exposure to chemicals* will be considered, with special attention given to plastics and rubbers, since we use them extensively in almost all aspects of our everyday life and it is very important to understand their part as chemical health hazards.

Some general facts concerning chemicals hazardous to health will be briefly presented here, followed by plastics and rubbers and related health hazards, in the next chapter. The reader who does not require a general introduction to hazardous chemicals may therefore skip this section and proceed directly to Chapter 3.

2.2.1 Chemicals that Cause Health Hazards

There are three main groups of chemicals to consider, with regard to health issues:

(i)   Toxic compounds,

(ii) Carcinogenics,

(iii) Endocrine disruptors.

2.2.1.1 Toxic Compounds, Their Toxicity and Exposure

Let us begin with a question: *What is a toxic compound?*

and with several definitions:

Toxicants are certain chemical (or physical) agents that have harmful or adverse effects on living organisms and can seriously damage or disrupt biological functions [5].

The word '**toxic**' is considered as synonymous with 'harmful' and 'poisonous', with regard to the effects of these chemicals [6]. However, although 'toxicity', simply, is 'the ability of a substance to cause injury to biological tissue', the hazard (or risk) posed by a substance is the 'probability that this substance can cause injury in a given environment', which is broader in definition. The hazard posed by any substance is characterised by factors such as its toxicity, the details of its absorption and metabolisation as well as excretion, its speed of action, perceptible signs of its possible hazard(s), and its potential (if there is any) for fire or explosion.

The word 'toxin' refers to a poisonous substance produced by a living organism, e.g., a bacteria.

Toxicology is the study of poisonous substances (chemical and physical agents) and their effects on living organisms.

It should be noted that, a hazardous chemical or a physical agent may be poisonous under certain conditions but harmless under others, the dose being one of the most important factors.

'Almost all chemicals are toxic at a sufficient dosage', in principle (discussed below). It is even possible to talk about 'the toxicity of water' [7] in this respect, so that all chemicals, at certain concentrations, are toxic.

At this point, it should also be noted that, since it is ethically impossible to test substance toxicities on humans, these are validated approximately with reference to animal experiments; however, the possible effects of differences between species are a fundamental problem, and hence such validation cannot be absolute and will not necessarily predict what will happen in man [8].

There are a number of different factors that can determine toxicity, including:

(i)   *toxicity of the chemical* (the chemical's inherent capacity),

(ii) *physical or chemical form of the agent* (whether it is a gas, vapour or particulate matter, such as an aerosol) and *the amount* of chemical/physical agent present (dose),

(iii) *route* of exposure,

(iv) *fate of the chemical in the body* after exposure (i.e., possible interactions or joint effects foreseen with other chemicals existing in the organism),

(v)   *differences in metabolism* between different species.

(i) 'Dose' is the 'number one' factor in toxic effect determinations

Paracelsus (1493-1541) once said: '*dosis facit venenum*', meaning 'toxicity is determined by the amount'. For example, the heavy metal copper is an essential nutrient required by the body and cannot be considered as a poison at low concentrations, and yet it *is* a poison at higher concentrations, causing lysis of red blood cells, vomiting and diarrhoea. Similarly, common table salt is an essential ingredient for life, but becomes quite toxic at high levels of intake.

A poison is defined as *a chemical that requires a dose of less than 50 mg per kilogram of body weight to kill 50% of the victims exposed,* which is about 3/4 of a teaspoon for the average adult and about 1/8 of a teaspoon for a 2-year-old child. Fortunately, there are few chemicals that are lethal at these doses.

Poisons can be acute (with immediate effect, e.g., hydrogen cyanide (HCN)) or chronic (referring to the systemic damage done after repeated exposure to low concentrations over long periods of time, e.g., heavy metals like mercury, lead, cadmium and also vinyl chloride). The chemicals most often associated with chronic toxicity are also carcinogens (e.g., benzene, cadmium compounds), which are problematic because when, if at all, the effect of the exposure will be felt cannot be estimated. Chemicals that were not thought to be hazardous in the recent past can found to be carcinogenic at a later time.

Most chemicals exhibit some degree of both acute and chronic toxicity. The symptoms and the systemic effects, however, will differ. Materials may act as acutely toxic substances, without showing any chronic effects, or *vice versa*. Despite this correlation, the effects of both forms of toxicity are always dose related: the greater the dose, the greater the effect.

The strength (or potency) of poisons is most frequently measured by the **lethal dose**. From statistical *dose to response* data, the dose (in mg per kg of body weight) killing 50% of a sample population is designated as the *median lethal dose* or *lethal concentration 50* (MLD or LD_{50}) (please see Appendix 2.A.2 for lethal dose/concentration and for toxic dose/concentration definitions). However, LD_{50} values may not accurately reflect the full spectrum of toxicity or hazard all the time, because some chemicals with low acute toxicity may have other harmful effects (e.g., carcinogenic or endocrine effects) even at very low doses that produce no evidence of acute toxicity at all. Usually the dose-response tests are done with animals and the results obtained are applied to humans with a 100-fold margin of safety (that is, if a test animal has a threshold of 100 ppm, the FDA will set the safe level for humans at 1 ppm, by assuming that humans are ten times as sensitive to the material as animals, and that the weak portion of the population (the old, ill and predisposed) is some ten times as sensitive as the healthy population).

(ii) Exposure and route of exposure

Exposure to chemicals is classified according to frequency and duration, and can be one of four types:

-   Acute exposure (exposure up to 24 hours),

-   Subacute exposure (repeated exposure for 1 month or less),

-   Subchronic exposure (repeated exposure for 1-3 months),

-   Chronic exposure (repeated exposure that lasts more than 3 months, often for 24 months or even a lifetime).

Route of exposure (which is either through inhalation, ingestion or skin contact) is another significant factor that can influence the toxic effect of a specific chemical.

Toxicity varies with *the route of exposure* as well as *the effectiveness with which the material is absorbed*. A chemical that enters the body in large quantities but is not easily absorbed poses a much lower risk than one that is absorbed easily into the bloodstream.

In general, there are five main routes of exposure for entry of a toxicant into the target organism:

(i)   *Via* the alimentary tract (through mouth and digestion system),

(ii) *Via* the respiratory system (inhalation exposure),

(iii) By percutaneous (skin) and perocular (eyes) absorption, and

(iv) By parenteral exposure (by injection),

and there may also be cases where a combination of these occur.

Substances are *absorbed* by the human body most efficiently through the lungs (inhalation), by the skin (contact) or by direct ingestion.

(ii.a) Absorption of toxicons 'by inhalation'

Toxic gaseous chemicals (called gaseous *toxicons*, after Paracelsus) are inhaled whenever toxic gases or vapours of VOC exist in the environment (gas and VOC vapour molecules differ in size: 0.0005 μm for gas and 0.005 μm for vapour). VOC vapour is mostly released by outgassing.

There may be also hazardous smoke or fumes existing in the air (with particles of airborne dust from 0.1 to 30 μm, of smoke from 0.01 to 1 μm and of fumes from 0.01 to 1 μm) [9].

The absorption of toxicons via inhalation by the target organism occurs first in the nose (upper airways), and then in the lungs (lower airways). The nose, for this route, acts as a 'chemical scrubber' (for water soluble and highly reactive gases, e.g., formaldehyde). Hence, highly water-soluble gases like ammonia, hydrogen chloride, etc., can dissolve in the moisture on the mucous coating, causing irritation. Substances with intermediate solubilities in water (like chlorine) can cause irritation at all points in the respiratory tract, while some insolubles (like nitrogen dioxide and phosgene) can also reach blood vessels rather easily.

Since the lung membrane is not an effective barrier, inhalation provides the most dangerous route of entry.

When gas molecules reach the lower airways, they can diffuse quickly within three-quarters of a second through the lung membrane and into the capillary network in the lungs and dissolve into the blood, to be carried to the rest of the body.

The rate of absorption of the toxic substances depends on their concentration, their solubilities in water, the depth of respiration and the rate of blood circulation. Because of the delicate nature of the respiratory system and blood vessels, an 'inflammatory response' occurs throughout, and many lesions are diagnosed as bronchiolitis etc., with 'acute, subacute or chronic' results depending on the stage of the response.

In the case of toxic particulates in air, particles with larger sizes (aerodynamic diameters of 5-30 μm) are caught in the upper airways, while particles with smaller sizes, of 1-5 μm and under 1 μm, can penetrate into the lower airways and blood vessels, respectively.

(ii.b) Absorption of toxicons 'through the skin (dermal)' and 'through the eyes' (percutaneous and perocular absorption)

Some toxic substances can be absorbed by the skin through its permeable texture and can then be distributed quickly into the bloodstream. This is the most common route of exposure. Since skin structure varies from the delicate (i.e., the scrotum with a thin keratin layer and high permeability) to rough (i.e., the soles, with a thick keratin layer and lower permeability), its permeability can differ similarly.

In the case of chemical contact with the skin, there is a range of possible interactions:

a) the skin acts as a barrier and the chemical cannot penetrate,

b) the chemical reacts with the surface of skin causing primary irritation (as is the case with acids, bases and a number of organic solvents),

c) the chemical penetrates the skin causing 'allergic contact dermatitis' (as seen with formaldehyde, and phthalic anhydride),

d) the chemical penetrates the skin and enters the blood (as in the case of aniline).

Skin usually acts as an effective barrier against the entry of most chemicals (i.e., inorganics), however, cuts and other abrasions can accelerate any absorption process. Depending on conditions, absorption of organic chemicals may or may not be realised easily as outlined above; even some organic chemicals can enhance absorption of others through the skin (e.g., dimethyl sulfoxide (DMSO)).

It has been shown that the eyes are also very effective 'open doors' for the absorption of toxicons into the body.

(ii.c) Absorption of toxicons 'by ingestion' (through the mouth and digestive system)

Chemicals that are ingested may be absorbed into the bloodstream anywhere along the gastrointestinal tract. If the material ingested cannot be absorbed by the body, it will most probably somehow be eliminated from the system.

Since we are considering here the toxic effects of plastic materials on humans in their everyday life; our main interest will be in the two exposure types mentioned above, which are the most common, namely, 'exposure *via* the *respiratory tract (inhalation exposure)*' and 'exposure by *percutaneous absorption (through the skin)* and *perocular absorption (through the eyes)*.

(iii) The fate of a chemical in the organism after exposure

The fate of a chemical, after an organism is exposed to it, is also an important factor in determining its toxic effects. It involves all possible interactions or joint effects of the chemical with other chemicals already existing in the organism. In this context, the water solubility, tissue reactivity and blood-to-gas-phase partition coefficient values of toxicons, are all important with regard to the VOC.

Many toxic substances that are stored in the body, mainly in fat or bone, can keep circulating throughout the organism for a long time.

2.2.1.2 Differences in Metabolism

The differences in metabolism between different species is usually another factor that determines toxicity and the extent of toxic effects.

In fact, the differences in metabolism and physiological conditions between different species may result in some chemicals being harmless to certain vertebrates, but toxic or highly toxic to others (this made possible the development of selective pesticides, which can kill chosen pests selectively while being relatively harmless to the operator and to other animals likely to be in contact with them, e.g., norbormide, a heterocyclic nitrogen compound which is extremely poisonous to rats but has negligibly low toxicity to humans and domestic animals).

Although the sensitivity to chemicals may vary from individual to individual, the target function or organ does *not* vary. One should also remember that the animal experiments that provide reference toxicities do not always give absolute results, if extrapolated to humans [8]: *The results obtained with animals do not necessarily predict that the same should happen with man (**the toxicogenomic approach**)*.

When a chemical comes into contact with a vertebrate, there may be:

(i) *Local (or topical) effect*, which is the direct corrosive effect of the chemical damaging tissues on contact,

(ii) *Systemic effect*, which is the effect after absorption into the organism, or

(iii) *Both of these effects*: e.g., caustic acids and alkalies, when ingested, will have a direct corrosive effect (on the lips, buccal cavity etc.), in addition to a systemic effect as they pass down the alimentary tract.

Systemic toxic effects will our main interest throughout this book.

Systemic toxic effects can be classified into the following five general groups:

a) *Independent Effect*: In this case, a combination of different toxic substances exists and each different chemical exerts its own effect independently of the others,

b) *Additive Effect*: This is the case when chemicals with similar toxicities produce a response equal to the sum of the effects produced by the individual substances,

c) *Antagonistic Effect*: This is the case when chemicals oppose or interfere with each other's toxicity,

d) *Potentiating Effect*: In this case one chemical enhances the toxicity of another in the body,

e) *Synergistic Effect*: This is when two chemicals produce a toxic effect greater than their sum.

One should also point out the differences between two types of toxicity:

Acute toxicity (effects occur immediately, shortly after a single exposure), and

Chronic, or subacute toxicity (delayed systemic effects that occur after long-duration repeated exposures to a chemical).

An acutely toxic substance can cause damage as the result of a single and short-duration exposure. In general, however, it may also have an effect after a long latency period following the single exposure in some rare cases.

The effects of both forms of toxicity are dose related, hence, the greater the dose, the greater the effect is. It should also be noted that, some acutely toxic substances may not show chronic ill effects, and certain chronic toxicons may not have any adverse single dose effect.

The materials most often associated with chronic toxicity are carcinogens (or carcinogenics).

Durations of short (up to 24 hours) and long-term time spans (in excess of 24 hours) must be specified [7]. *Hence, multiple or continuous exposures for up to 24 hours are considered as acute toxic exposure.*

Individuals with certain health problems (such as diseases of the liver or lungs) are likely to be affected more by exposure to toxic substances, and once exposed, to experience a more severe reaction.

2.2.1.3 Detoxification

There may be different possible routes to consider for toxicons in the body after the exposure, and **detoxification** is one of them. Detoxification occurs whenever the chemical absorbed is altered or metabolised somewhat (either by breaking down into products that can be incorporated or excreted, or by producing less toxic metabolites).

The chemical absorbed and its metabolites that are *excreted* or *stored* or *transported* in the organism may reach sites where toxic effects can be induced (e.g., they may concentrate in a specific tissue, such as in the liver, kidney, adipose tissue, etc).

If excretion is rapid, the effect is usually of low toxicity. If excretion is slow, there is the potential for more serious long-term effects. In the case of the complete excretion of an initial dose, successive intakes of the same doses are also excreted, as long as no residue remains. Otherwise, it is possible for the residue of the second dose to add to the first and, if doses are repeated often enough, to reach a toxic concentration.

Some notes to consider:

General information and updates about toxicons are available through a number of websites, e.g., the National Toxicology Program (NTP) of the US, National Institute of Environmental Health Sciences (NIH-DHHS) [10].

2.2.2 Carcinogen(ic)s

A carcinogen(ic) is an external substance or agent capable of inducing various un-regulated growth processes in cells or in tissues to give rise to malignant neoplasms that cause cancer, which is one of the three leading causes of death among humans.

A chemical can be considered as a carcinogen if it has been evaluated by the International Agency for Research on Cancer (IARC) of the World Health Organization (WHO) and found to be a carcinogen, or a potential carcinogen, and/or it is listed in the *Annual Report on Carcinogens* published by the National Toxicology Program (NTP) and/or it is regarded by the Occupational Safety and Health Administration (OSHA) as a carcinogen.

There are a number of *suspected* and *proven* cancer-causing chemicals, or *external environmental factors*, which account for most of the causes of cancer (about 80-90%), e.g., the tar present in cigarette smoke. There are also a number of *internal factors* that can cause cancer, such as irradiation (radioactive, ultraviolet and radiofrequency), certain viruses (such as polyoma), and diminished immune functions due to heredity, old age, poor state of health, etc.

Carcinogens are considered as a special class of chronic poisons and hence the subject of a specialised field of toxicology.

Because of the different factors involved, not everyone necessarily responds in the same way to the same external and internal factors, as regards the cancer issue.

Carcinogens, such as organic and inorganic chemicals with various biological actions, can act similarly to other toxic substances, with similar dose-response relationships but several distinct differences.

In medicine, a carcinoma is any cancer that arises from epithelial cells. It is malignant by definition: carcinomas invade surrounding tissues and organs, and may spread to lymph nodes and distal sites (metastasis). *Carcinoma in situ* (CIS) is a pre-malignant condition, in which cytological signs of malignancy are present, but there is no histological evidence of invasion through the epithelial basement membrane.

According to a report from WHO [11], 35% of carcinogenic substances are derived from chemicals connected to food and drink, and some 30% are from smoking (from the tar) [7]. There is a long active list of known and suspected chemical carcinogenic substances, classified as likely or probable [11, 12, 13, 14, 15], which is renewed periodically, containing expected carcinogenic chemicals and some unexpected (such as some tranquillisers and antibiotics, antipyretics, analgesics etc., that may lead to malfunction of the liver, and ultimately may cause liver cancer). Substances are usually assigned the risk numbers, e.g., R45 (may cause cancer in general) and R49 (may cause cancer by inhalation).

There are also claims that some anti-cancer medicines can themselves induce carcinomatosis [7].

It is proven that a number of agricultural chemicals are also carcinogenic in nature [16].

Cancer was first identified in the late eighteenth century, after observation of its incidence in patients who were chimney sweepers in the UK, which established the interrelation between chemicals and the disease (exposure to soot, coal tars and benzene). Later, the carcinogenic potency of tar was shown to be related to its *polynuclear aromatic hydrocarbon structure*.

Based on chemical and biological properties, carcinogens can be separated into two general classes:

1. *DNA-reactive carcinogens*: the most common human carcinogens. They are active with a single dose, and often such toxic effects are cumulative. They can act synergistically with one another.

2. *Epigenetic (EDC) carcinogens*: 'genotoxic' carcinogens. They are not DNA reactive and appear to operate by the production of other biological effects.

Plastics, rubber and asbestos are considered to be in the EDC carcinogenics group.

Three agencies are mainly responsible for evaluating data on carcinogenicity: *the IARC of WHO, the NTP and OSHA*. These agencies each perform very different functions in determining carcinogenicity, analysing the results, and making recommendations.

The IARC classifies carcinogenic chemicals into the following three categories:

a) *Group 1* (**carcinogenic** to humans),

b) *Group 2A* (**probably carcinogenic** to humans; when there is limited evidence of carcinogenicity in humans but sufficient evidence in experimental animals), and

c) *Group 2B* (**possibly carcinogenic** to humans; when there is limited evidence of carcinogenicity in humans and no or inadequate supporting evidence in experimental animals).

The NTP prepares periodical reports on carcinogens as 'known carcinogens' or 'reasonably anticipated to be carcinogens', while OSHA regulates a number of specific carcinogenic materials through the standards issued.

Table 2.1 lists some well-known toxic substances and carcinogenic agents.

Table 2.1 Some common toxic substances and carcinogens in humans		
Common Acute Poisons	**Common Chronic Poisons**	**Common Carcinogen(ic)s**
Cyanides and nitriles Bromine Chlorine Fluorine Iodine Heavy metals Hydrogen cyanide Hydrogen sulfide Nitrogen dioxide	Heavy metals Mercury Lead Vinyl chloride	Acylating agents Alkylating agents Alpha-halo ethers Sulfonates Epoxides Electrophilic alkenes and alkynes) Aromatic amines Aromatic hydrocarbons benzene Carbon tetrachloride Cadmium compounds Hydrazines N-nitroso compounds Organo-halogen compounds

Several related web sites are listed in references [17, 18, 19, 20].

2.2.3 Endocrine Disrupters (ECD)

2.2.3.1 The Endocrine System and ECD

The endocrine (hormonal) system is a network of glands and receptors that function to regulate a number of key body-functions, like growth, development, and maturation, as well as the operation of various organs in the bodies of vertebrates. In short, *the endocrine system provides the key communication and control link between the nervous system and all body functions*, through the secretion of hormones.

There are a number of chemicals that can disrupt normal functioning of the endocrine system, called endocrine disrupters (ECD). *An ECD usually either mimics or blocks hormones and hence disrupts the body's normal functions, after its absorption by the body.*

Most ECD are synthetic chemicals that cause 'hormone-related' diseases, mainly connected to 'reproduction' and 'dysfunction' at very low levels (even at parts per trillion). Environmental exposure to any of these chemicals usually results in disruption of thyroid hormones, androgens, oestrogens and other endocrine processes in human and wildlife organisms. Synthetic oestrogens are produced either through industrial manufacture or as by-products of such processes or by burning. Known synthetic oestrogens have been identified by laboratory tests such as those that measure a chemical's ability to speed the growth of cultures of breast cancer cells. The mechanisms of ECD is poorly understood and specific endpoints or effects of ECD are not yet completely defined, and there is still much to be understood and to be explored about its role.

The effects of ECD on **developing organisms** are of greatest interest, because the disruptive (organisational) effects in this case are shown to be permanent and irreversible, while ECD exposure of **adults** can be reversible [21].

A wide range of organic chemicals (mostly synthetics, *including certain additives and plasticisers* - 'such as phthalates, bisphenol A, and nonylphenols'- *all used as additives in plastics*) are suspected potential ECD agents. The list of ECD also includes pesticides (such as DDT, which is banned) and many industrial and consumer products - such as, liquid soaps, shampoos, conditioners, and hair colours - that contain alkylphenol ethoxylates (APE, which have been replaced by the more expensive, but much safer, alcohol ethoxylates). In addition, there are polychlorinated biphenyls (PCB), dioxins, certain preservatives and metal ions, and even certain treated woods suspected as endocrine disruption agents in humans based on (limited) animal studies, (see **Table 2.2**). Synthetic oestrogens are the focus of current concern for ECD; there are certain natural 'oestrogens', i.e., 'phytoestrogens', which are much safer. They occur in a variety of plants in nature and can be safely metabolised (or are degradable), and so do not bioaccumulate in living organisms.

Disruption of the endocrine system is believed to occur in the following ways:

a)   Some ECD can *mimic* a natural hormone, hence fooling the body into over-responding to the stimulus (e.g., a growth hormone that results in increased muscle mass), or responding at inappropriate times (e.g., producing insulin when it is not needed).

b)   Some ECD can completely block the effects of a hormone on certain receptors (e.g., growth hormones required for normal development).

c)   Some ECD can directly stimulate or inhibit the endocrine system and can cause overproduction or underproduction of hormones (e.g. an over- or under-active thyroid).

None of these effects are desirable. However, ECD can also sometimes be helpful in humans: certain drugs such as birth control pills, although known ECD, are widely used.

ECD are of particular interest to the environmental and medical sciences, but although there is some data for the ECD effect of pesticides, ECD – related risks for most of the chemicals (approximately 90,000) are still not available, and many scientific uncertainties exist.

Several examples can be given of the effects of ECD. The case of the synthetic oestrogen, diethylstilboestrol (DES), used as a potent drug in the recent past, is usually taken as one such example. DES is banned now, but in the early 1970s, doctors prescribed DES to about five million pregnant women, to block spontaneous abortion or to promote foetal growth. In fact DES affects the development of the reproductive system and can cause (vaginal) cancer, as understood later. Following such examples, there is now the recent requirement that any new application is subject to the proper ECD screening programs

In another case, workers engaged in artificial leather manufacture had been exposed to dimethylformamide (DMFA) for between one and five years, which was found to have had an adverse effect on pregnancy and childbirth; increasing birth complications [17].

In another study, it is shown that exposure to sStyrene monomer is associated with serious menstrual disturbances and additional diseases: *chronic illness with secondary amenorrhea, nulliparity with both decreased blood clots and hypermenorrhea.*

In 1996, the Environmental Protection Agency's (EPA) Office of Research and Development (ORD) [15] identified *endocrine disruption as one of its top six research priorities*. ORD's research program plan published in 1998, has three long-term goals in ECD research [11].

The EU adopted 'the Community Strategy for Endocrine Disrupters' in 1999 to focus on short, medium and long-term actions [22], and the EU Commission has already finalised a study on *Information Exchange and International Coordination on Endocrine Disrupters*, through the MRC Institute for Environment and Health (UK) [23]. Under existing legislation within the EU, the assessment of the potential toxicity of a chemical is dependent mainly on the type of the chemical, and different chemicals are subject to different legislation governing their testing and assessment.

In general, synthetic chemicals prior to marketing will undergo testing (hazard identification) and the results of this will be used in the risk assessment (second level), after which there is risk management (as the third level, to determine whether any restrictions are needed for use).

Existing EU legislation accounts for ECD effects on reproduction and related diseases, such as cancer, but not on any other disruption to the endocrine system.

2.2.3.2 Polychlorinated Biphenyls (PCB) and Polychlorinated Dioxins (PCD)

Polychlorinated biphenyls (PCB), are a family of toxic, oily, non-flammable chemicals. They are man-made products and were first commercialised in 1929 (by Monsanto). They were mainly used in electrical equipment (e.g., transformers and capacitors), as heat transfer and hydraulic fluids, and as plasticisers from World War I until recently, mainly due to their exceptional thermal and chemical stabilities. Although their production in the USA was stopped in 1977 (they were banned worldwide), some production still continues, and it is believed that large quantities of PCB may still be present in some old transformer and capacitor systems. PCB are certainly still present in the USA in some electrical equipment and are frequently found at toxic waste sites and in contaminated sediments worldwide. The sealants based on polysulfhide polymers that were used in buildings some 20-40 years ago contained PCB, which has been shown still to exist at alarming levels in some houses in Sweden, [24].

It has been shown that children exposed to low levels of PCB in the womb, through the mother's consumption of contaminated fish, grow up with low IQ, poor reading comprehension and with memory problems.

As shown in **Table 2.2**, at the present time, the main sources of PCB are industrial chemical production or industrial by-products, from landfill, and incinerators.

Table 2.2 Some of the probable sources for ECD	
ECD	Source(s)
(Some) Additives Brominated flame retardants	Plastics and textiles
Alkylphenols Nonylphenol	Surfactants (certain detergents and their metabolites)
(Natural) Hormones	Animals
(Some) monomers Bisphenol A	Lacquers used in dental treatment, internal coatings for metal containers such as food cans
Organochlorine Pesticides DDT Dieldrin Atrazine (Other) Pesticides	Mostly phased out Herbicides, insecticides, fungicides
Organotin Compounds Tributyltin	Antifouling paints for ships
Polychlorinated Compounds Polychlorinated dioxins (PCD) Polychlorinated biphenyls (PCB)	Industrial production or by-products, landfills, incinerators.
(Some) Phthalates Dibutyl Butylbenzyl phthalates	Plasticisers used with plastics
Phytoestrogens Isoflavones Lignans	Pulp mill effluents
(Some) preservatives Parabens	Cosmetics, some antibacterial toothpastes (mimicking oestrogens)
(Synthetic) Steroids	Contraceptives

Dioxins are a family of the most toxic chlorinated organic compounds known to science, numbering around 75 dioxins and 135 related furans. These can cause cancer and are ECD for humans, even at very low exposure levels, since minute amounts, can bio-accumulate due to their ease of solubility in body fat (dioxins are *hydrophobic*, 'water-hating' and *lipophilic*, 'fat-loving'). Number and position of chlorine atoms in the molecule has a considerable effect on toxicity, and **17 dioxins are classed as *highly toxic*.** These include polychlorinated dioxins (PCDD) and dibenzofurans (PCDF) which are by-products of the chlorine bleaching of paper, the burning of chlorinated hydrocarbons (such as pentachlorophenol, PCB, and PVC) and the incineration of municipal/medical wastes, and also from natural events (e.g., from forest fires, traffic exhaust and even volcanic eruptions). These toxic substances can easily contaminate the soil, and can bio-accumulate in fish (to the extent that dioxin levels in fish are usually 100,000 times that of the surrounding environment) and in other wildlife. *Hence for humans, the most common route of exposure to dioxins is through the food chain.* It is a fact that livestock as well as wildlife and humans are all presently exposed to dioxin-like compounds at different levels, and after considering the incident in Yusho (1968, Japan) and Yuchen (Taiwan, 1979), the use of so-called Agent Orange in Southeast Asia (Vietnam, 1962-1971), and the accident in Seveso (Milan, Italy, 1976), it is believed that '*the general population of the industrialised world carries considerable quantities of these toxins in their bodies*' [2, 25, 26].

Most PCB are believed to have dioxin-like toxicities, that is, coplanar PCB with four or more chlorines and with one or no substitution in the *ortho* position, generated and released by combustion and incineration processes when PCB is present. One should note that *para* or *meta* substitutions lead to highly toxic planar compounds [27].

The International Agency for Research on Cancer (IARC), which is a part of 'WHO' has classified the most common and toxic dioxin 2,3,7,8-tetrachlorodibenzo-*p*-dioxin (2,3,7,8-TCDD) as a *known 'Class 1' human carcinogen* [28].

Persistent organic pollutants (POP) are certain organic compounds (mostly polycyclic aromatic hydrocarbons) that are directly carcinogenic or can be metabolised to carcinogens), and which at minute concentrations have the potential to damage human health, through inhalation or through ingestion with food (including in the vapour phase or by absorption depending on temperature. Dioxins and PCB are well known POP. All POP are subject to the Stockholm Convention, which obliges signatory countries to take the necessary measures to eliminate (whenever possible), or to minimise (where elimination is not possible) all sources of dioxins [29, 30, 31] The decrease is even projected to reach 'zero discharge' by the year 2050 [32].

A more comprehensive information on the general issue of toxicity for plastics, rubber and composite materials is provided in Chapters 3, 4, 6 and 10.

2.3 A Final Note

Revisions are being made to chemical control measures by describing the key facts and objectives of future chemicals policy in EU legislation through the Registration, Evaluation and Authorisation of Chemicals (REACH) regulation, which will be effective in 2007. REACH is expected to have major implications for European chemical producers and downstream users, since it is anticipated that around 30,000 chemicals will need to be screened for their health and environmental impact [8, 33, 34, 35].

References

1.   G. Graff, *How to Keep Plastic Odours under Control*, Omnexus Trend Report, 4 May 2005, www.omnexus.com

2.   M.J.M. Brown, J.L. Licker and M.R. Zbuchalski, inventors; International Flavors & Fragrances, Inc., assignee; US 20050129812 A1, 2005.

3.   A.S. Yuwono and P. Schulze Lammers, *Agricultural Engineering International: the CIGR Journal of Scientific Research and Development*, 2004, **VI** (Invited Paper).

4.   C. Henneuse Boxus and T. Pacary, *Emissions from Plastics*, 2003, Rapra Review Reports, **14**, No.161.

5.   M.O Amadur, J. Doull and C.D.K Klaassen, *Csarett and Doull's Toxicology - The Basic Science of Poisons*, 4th Edition, Pergamon Press, New York, USA, 1991.

6.   T.A. Loomis and A.W. Hayes, *Loomis's Essentials of Toxicology*, 4th Edition, Academic Press, New York, USA, 1996.

7.   V. K. Brown, *Acute and Sub-Acute Toxicology*, Edward Arnold, London, UK, 1988.

8.   *RTD Info*, 2006, **48**, 34.

9.   H. McDermott, *Air Monitoring for Toxic Exposures*, 2nd Edition, Wiley Interscience, Hoboken, NJ, USA, 2004.

10.   NTP: ntp.niehs.nih.gov; NLH ChemID Plus Advanced: chem.sis.nlh.gov/chemidplus

11.   World Health Organisation(WHO): www.who.int/en; www.aboutcancer.info/Carcinogenics/carcinogenics.html

12.   *List of Substances which are Carcinogenic, Mutagenic or Toxic to Reproduction etc*, University of Bristol, www.chm.bris.ac.uk/safety/Carcinogenetclist.htm

13.   Environmental Protection Agency (EPA):
www.epa.gov/ttn/atw/nata/34poll.html
www.epa.gov/ORD/WebPubs/final
www.epa.gov/ord/index.htm

14.   *Acronym: List of Carcinogenic Substances (Sweden)*, University of Kassel, dino.wiz.uni-kassel.de/dain/ddb/x339.html

15. *Control of Substances Hazardous to Health (COSHH)*, Heath & Safety Executive, www.hse.gov.uk/coshh/

16. *Genetically Manipulated Food News, 13 January 99*, The Safe-Food-Coalition of South Africa, home.intekom.com/tm_info/rw90113.htm

17. European Environmental Agency (EEA): www.eea.int glossary.eea.eu.int/EEAGlossary

18. Canadian Centre for Occupational Health and Safety (CCOHS), www.ccohs.ca/oshanswers/chemicals/endocrine.html

19. The American Conference of Governmental Industrial Hygienists (ACGIH), www.acgih.org

20. The American Industrial Hygiene Association (AIHA), www.aiha.org

21. J.P. Myers, L.J Guillette Jr, P. Palanza, S. Parmigiani, S.H. Swan and F.S. Vom Saal, *International Seminar on Nuclear War and Planetary Emergencies – 30th Session*, Erice, Italy, 2003, 105.

22. *Community Strategy for Endocrine Disruptors* (EU-COM 706 1999), European Commission: http://europa.eu.int/comm/environment/endocrine/index _en.htm http://ec.europa.eu/environment/endocrine/documents/studies_en.htm

23. *Endocrine Disrupting Chemicals*, Eds., R.E. Hester and R.M. Harrison, Royal Society of Chemistry, Cambridge, UK, 1999.

24. *ENDS Report*, 1997, **266**, 11.

25. T. Webster and B. Commoner in *Dioxins and Health,* 2nd Edition, Eds., A. Schecter and T. Gasiewicz, Wiley Interscience, Hoboken, NJ, USA, 2003, p.1.

26. A. Schecter in *Biological Basis for Risk Assessment of Dioxins and Related Compounds* (Banbury Report 35), Eds., M.A, Gallo, R.J, Scheuplein and K.A, Van der Heijden, Cold Spring Harbor Laboratory Press, Plainview, NY, USA, 1991, p.169.

27. J-Y. Wu, K. Pan and T-I. Ho in *Environmental Applications of Ionizing Radiation*, Eds., W.J. Cooper, R.D. Curry and K.E. O'Shea, John Wiley and Sons, New York, USA, 1998, p.283.

28. *Polychlorinated Dibenzo-para-Dioxins and Polychlorinated Dibenzofurans*, IARC Monographs on the Evaluation of Carcinogenic Risks to Humans: Volume 69, IARC, Lyon, France, 1997.

29. N.Y. Ivanova and S.A. Serednitskaya, *Gigiena Truda i Professional'nye Zabolevaniya*, 1989, **7**, 28.

30. G.K. Lemaster; A. Hagen and S.J Samuels, *Journal of Occupational Medicine*, 1985, **27**, 7, 490.

31. *Dioxin Homepage*, ActionPA, www.ejnet.org/dioxin

32. T.E. Graedel in *Handbook of Green Chemistry and Technology*, Eds., J.H. Clark and D.J. Macquarrie, Blackwell Publishing, Oxford, UK, 2002, p.56.

33. R.A. Kerr, *Science*, 2000, **289**, 5477, 237.

34. *Profile of the Petroleum Refining Industry* (EPA/310-R-95-013), US Environmental Protection Agency, Washington, DC, USA, 2000.

35. *RTD Info*, 2006, **48**, 33.

Appendix 2.A.1 Some Organic Indoor Pollutant Classifications by WHO

The World Health Organisation (WHO) classifies organic indoor pollutants according to their boiling points (bp), as follows:

a) The most common and critical organic pollutants, with regard to plastics materials, are volatile organic compounds (VOC) with bp between 50 °C and 260 °C at ambient.

b) Organic pollutants with bp from 0 °C to 100 °C are very volatile organic compounds (VVOC).

c) Organic pollutants with high bp from 240 °C to 400 °C are semi-volatile organic compounds (SVOC).

d) Organic pollutants with much higher bp above 380 °C are particulate organic compounds.

The VVOC and VOC are mostly due to non-bound, rather low molecular-weight organic molecules (also called as *free* or *primary emissions*) that exist in the system.

However, there are cases where bound organic molecules, or some parts of the system itself, can also contribute to VOC and VVOC emissions, if there is any special effect that can break them down from the system (e.g., thermal, chemical or mechanical degradation), and these are termed *secondary emissions*.

VOCs are mainly hydrocarbons (2.73 Mt in 1997 in the UK) and the *most abundant is methane* (50% from landfills, 30% from animals, and rest from gas extraction, biomass burning) while *non-methane* VOC in 1997 in the UK (2.13 Mt in the UK) are mainly emitted from vehicles (40%), from solvents in paints (30%), etc., which is composed of some 200 different hydrocarbons such as benzene and toluene [34].

An EPA Report in 2000 claimed industrial emissions of VOC had declined by over 40% since 1995 in the UK [35].

Appendix 2.A.2 Some Definitions of Lethal and Toxic Doses and Concentrations

The following specific definitions should be considered for lethal and toxic doses and concentrations:

LC_{50} is defined as the concentration of the toxicon **in air** that is expected to kill 50% of a sample population during a single exposure in a specified time period.

LC_{LO} is the lowest concentration of a toxicon **in air** causing death, in exposure periods less than (acute) or greater than (subacute and chronic) 24 hours.

LD_{50} is the single dose that causes the death of 50% of a sample population following exposure by any route **other than inhalation**.

LD_{LO} is the lowest dose of a toxicon that causes death following exposure by any route **other than inhalation**.

TC_{LO} is the lowest concentration of a toxicon **in air** that causes any toxic or tumorigenic or reproductive effect following exposure for any given period of time.

TD_{L} is the lowest dose of a toxicon that causes any toxic or tumorigenic or reproductive effect following exposure for any given period of time by any route **other than inhalation**.

Appendix 2.A.3 Inherent Toxicity Levels of Chemicals Hazardous to Health (OSHA)

Definition of 'Toxic' and 'Highly Toxic'

LD_{50} (oral/digestive system): Toxic if > 50-500 mg/kg; Highly Toxic if < 50 mg/kg.

LD_{50} (dermal exposure): Toxic if > 200-1000 mg/kg; Highly Toxic if < 200 mg/kg.

LC_{50} (inhalation): Toxic: if > 200-2000ppm (2-20 mg/l); Highly Toxic if < 200 ppm.

There are also several special systems that have been developed for grading chemical health hazards, e.g., by the National Fire Protection Association (NEPA), considering mainly chemicals evolved under fire conditions; which are rated from 0 (non-toxic) to 4 (extremely toxic).

Appendix 2.A.4 Some OSHA and ACGIH Definitions of Exposure Limits

The Occupational Safety and Health Organisation (OSHA, www.osha.gov), the National Institute of Occupational Safety (NIOSH, www.cdc.gov/niosh) and the American Conference of Governmental Industrial Hygienists (ACGIH) are the three main agencies that provide information on safe exposure limits for airborne contaminants, developed mainly for workers.

It should be noted that, in general, exposure levels in the work place are considered to be much higher than for the public at large.

(a) Some OSHA definitions

The OSHA definition of hazardous chemical:

A hazardous chemical is defined in accordance with the following four references:

i)   **US Code of Federal Regulations** (CFR) 29 part 1910 subpart Z) Toxic and Hazardous Substances.

ii) **International Agency for Research on Cancer (IARC),** Monographs on the Evaluation of Carcinogenic Risk of Chemicals to Humans.

iii) *ACGIH,* Threshold Limit Values for Chemical Substances and Physical Agents in the Work Environment.

iv) **National Toxicology Program (NTP),** Annual Report on Carcinogens.

(b) Other definitions:

AL *(Action Level)* is the exposure level at which the protective programme required by OSHA regulations must be put into effect. This would include things such as air monitoring, medical surveillance and training.

C *(Ceiling)* is the concentration of a substance that should not be exceeded.

PEL *(Permissible Exposure Limit)* is the maximum allowable exposure (in amount and time) to an airborne contaminant for a worker on a daily basis, to avoid suffering adverse affects.

*The PEL established by OSHA for each chemical are universally considered as permissible legal limits for exposure. In general, there are **two PEL values for each chemical**:*

(i) Time-Weighed Average Limit (TWA): the maximum average airborne concentration of contaminant acceptable (in ppm) over an eight-hour period, to which workers may be exposed for any eight-hour day of a 40-hour week. This level may not be appropriate for the old, young, ill or those predisposed to problems from chemical exposures.

(ii) Short-Term Exposure Limit (STEL): the maximum concentration to which exposure is permitted (in ppm) as averaged over a short time (e.g., a 15-minute period).

OSHA in alliance with the Society of Plastics Industry maintains a web page of information relevant to the plastics industry at www.osha.gov/SLTC/plastic/index.html.

(c) Some ACGIH Definitions

TLV *(Threshold Limit Value)* is the permitted airborne concentration of a substance to which nearly all workers may be exposed without adverse effects, for eight hours in each day as part of a 40-hour week.

(TLV were developed long before OSHA's PEL values. TLV are essentially the same as PEL - except that PEL carry the force of law whereas TLV are only recommendations.)

TLV-TWA *(Threshold Limit Value - 'Time Weighted Average')* is the allowable 8 hour-per-day concentration to which a worker may be exposed during a 40-hour week.

TLV-C *(Threshold Limit Value - 'Ceiling')* is the ceiling value that should not be exceeded at any instant. Unlike the other TLV, which serve as guidelines, the TLV-C must be viewed as an absolute boundary.

TLV-STEL *(Threshold Limit Value - 'Short-Term Exposure Limit')* is the maximum concentration of a substance which a worker may be exposed over a continuous 15 minute period with a low probability of experiencing irritation, irreversible damage, or unconsciousness. Four of such 15-minute periods are allowed per workday of 8 hours, with at least a 60-minute break between them. *However, at no time may the TLV-TWA be exceeded.* In the case of any exposure, which is above TWA yet below STEL should be clearly specified so. However, in any case such exposures should be 'for *no longer than 15 minutes at any one time or no more than 4 times per day*'.

3 General Issues of Toxicity for Plastics and Rubber

3.1 Plastics and Rubber, In Brief

Plastics are defined as shaped and hardened synthetic materials composed of long chain organic molecules called polymers, plus various additives. Hence, plastics are not pure, in what they contain, and they must also be shaped to their final form of use.

The physical and chemical properties of a plastic material are mainly determined by the size (molecular weight) and structure, respectively, of the polymer molecules of which it is composed.

Polymers are large molecules composed of two to several thousand simple molecules as repeating units, called *monomers*. Monomers are converted into the polymer through special reactions known as polymerisation. Hence the polymer is the pure material. Almost all synthetic polymers are synthesised from petroleum, although there are natural polymers as well, for example, cellulose, wool and so on.

Rubbers are highly elastic materials composed of polymers (because of their high elasticities, rubbers are considered to be a special class of plastics).

There are two main types of plastics, in terms of their response to heat:

Thermoset plastics, such as Bakelite, which stay hard once set and do not soften or melt with increasing temperature. At very high temperatures molecules in a thermoset decompose with evolution of gas, and after complete decomposition a solid residue remains. Hence, the response of thermosets to heat is irreversible.

Thermoplastics, such as polyethylene (PE), on the contrary, are solid at ambient temperature, but soften and transform into a highly viscous melt at higher temperatures. If the temperature of the thermoplastic melt is then decreased, it solidifies, and can be re-softened and melted again by heat. This cycle (of softening to melting if heated, and melt to solid if cooled) can be repeated indefinitely, and hence the response of thermoplastics to heat is, characteristically, reversible.

There is also the frequently-used term *resin*. Resins are synthetic or natural polymers that are liquid and sticky at room temperature (meaning, uncured in the case of the thermoset or of low molecular weight if thermoplastic).

Artificial resins include polyesters and epoxies, which are mostly used as adhesives and binders.

Natural resins are usually secreted by various plants (e.g., oleoresin).

There are over 30,000 different natural and synthetic polymers known today, and about 10% of these are synthetic, and the rest (about 27,000) are natural. Synthetics are man-made, and virtually all synthetic plastics and rubbers are derived from petroleum (crude oil), natural gas and coal as feedstocks. These feedstocks consume only 6-8% of oil and gas (4-5% is required for their direct production and their processing energy accounts for another 2-3%). Among the synthetic polymers, there are *commodity polymers* (such as PE, polystyrene (PS), polyvinylchloride (PVC) and polypropylene (PP) that are used in everyday life and hence are the most common. In addition, there are special polymers designed and produced for special purposes (e.g., high performance engineering polymers, characterised by their high tensile strengths greater than 40 MPa - exhibiting high stability performances in continuous use at temperatures above 100 °C). Among engineering polymers, are polyamide (PA), also called Nylon, polybutylene terephthalate (PBT), polyethylene terephthalate (PET), polyoxymethylene (POM), polycarbonate (PC) and polyphenyl ether (PPE), polyethyl sulfone (PES), polyetheretherketone (PEEK), and polyether-imide (PEI).

Annually more than 150 million tons of plastics (corresponding to about 20 kg per person) are produced worldwide. The greatest use (amounting to 40% in the EU) is in *packaging* which provides added protection (and prolongs freshness in the case of food) to the goods packaged. The second greatest use of plastics is in building and construction material and as consumer products. Transportation, electrical and electronic applications, agricultural, and biomedical/healthcare applications of plastics are also significant.

Thus, plastics have become a very important part of modern life, however, there still exist concerns over their possible effects on health and over associated ecological problems all of which need to be understood and resolved as favourably as possible.

3.1.1 Combinations of Plastics, Combinations of Rubbers

Until the 1960s, plastics and rubber were used mostly as consumer materials, with plastics used mainly in packaging applications. After this period, however, more sophisticated types of special polymer were evolved for different, demanding applications with combinations of plastics as well as rubbers.

These include:

a) *Composites* (consisting of polymers with fillers and/or fibres),

b) *Blends* (a mixture of different polymers), and

c) *Laminates* (consisting of layers of polymers).

3.2 Additives

Most polymers are of little value in their pure form because of poor physical properties, which must therefore be regulated properly with certain additives, both to facilitate handling and processing as well as to impart the desired properties to the final product. Hence, plastics and rubbers are composed of 'polymers' plus 'certain additives', the latter used for the purpose of changing or improving various (chemical, mechanical and physical) properties of the base polymer.

There are already over 4000 different types of additives available for plastics and rubbers in a global business of around $16 billion per year.

Additives are a complex group of certain chemical derivatives and minerals accounting for 15-20% by weight of total plastic products marketed. Additives play the key role in improving and creating the range of unique performance characteristics in plastics and protecting them from the effects of time, heat and environmental conditions. Usually, additives are *stabiliser systems* (to ensure durability) and *plasticisers* (to produce a range of flexibility), in addition to their other possible functions (i.e., antimicrobial, lubricating, pigmenting, flame retarding, impact modifying, anti-static and antioxidant, ultraviolet (UV) absorbing, compatibilising, adhesion promoting, anti-fogging, dispersive, filling and extending functions). The proper mixing of polymers with additives results in a compound, and this process of mixing is called *compounding*. The compound is then processed by the appropriate processing method chosen to end up with the final plastic object.

Additives can be thought of as having two different, direct origins, directly, (as *intentional additives* and *unintentional additives*). There may also be certain compounds in the system that are due to some of the polymerisation ingredients left after completion of the polymerisation reaction as impurities (specifically *remnants*), and a fourth type, also an impurity, that can arise from interactions and can spoil the system. All these are outlined next.

(i) *Intentional Additives*

Intentional additives are added to the system for a specific purpose, such as, increasing flexibility, increasing heat resistance and so on, where the quantities and chemical characteristics are well known.

Intentional additives can be classified further in accordance with their intended functions:

a) *Process additives:* which are lubricants, mould-release agents, blowing agents and so on,

b) *Stabilisers:* such as heat stabilisers, UV and visible light stabilisers, antioxidants, antimicrobials, fire and flame retardants, and so on, and

c) *Performance additives:* which are fillers, reinforcing agents, fibres, colouring agents, impact modifiers, antistatic agents and plasticisers.

More detailed information about intentional additives is given in Section 3.2.1.

(ii) *Unintentional Additives*

Unintentional additives are introduced into the system 'unintentionally', with little knowledge of their existence and characteristics. Impurities in the intentional additives are considered unintentional additives. Their chemical formulae, properties and concentrations are usually not known, and they usually exist in rather small quantities - these uncertainties can cause a severe analytical problem.

(iii) *Remnants (of Polymerisation Ingredients, Catalysts and so on)*

Remnants of the polymerisation process (e.g., monomers, oligomers, catalysts and so on) are possible impurities after the polymerisation is completed. Their concentrations are not known and are usually very low. However, since most of the catalyst systems contain heavy metal ions that are poisonous, as will be outlined shortly, and since some of the monomers can also pose serious health hazards, these additives should be considered as another problematic group.

(iv) *Interaction Chemicals (that are 'Produced in the System')*

Any of the additives in a plastic or rubber material, outlined above, can cause the evolution of other chemicals, indirectly, through specific chemical reactions (chemical interaction) between the additives and some of the chemicals existing in the material with which they are in contact. Interaction chemicals are products of those additives that migrate and interact, and they can be hazardous, while their formulae and characteristics, as well as concentrations, cannot easily be known, hence, they can raise another critical issue to consider, in particular in the case of food-packaging plastics and food-contact rubbers.

3.2.1 Migration of Additives

Additive molecules, in almost all cases, are much smaller in size than the associated polymer molecules and are mostly organic, hence, they can usually evaporate and go into the gas phase rather easily, if their boiling points are low, or they can migrate (leach) from the plastic or rubber into the contacting material. If they are toxic, their toxicity will also be transferred either into the vapour phase (and can be absorbed by humans through inhalation), or if they migrate into edible matter such as food with which they are in contact, their toxicity will be transferred to the food (and can be absorbed by ingestion). Since most chemicals, at sufficient concentration, are toxic to some extent, all of the additives, namely, the 'foreign bodies' in the plastic system, should be recorded and analysed, both qualitative and quantitatively.

There are on-going studies into bonding the additive to the polymer backbone to block and hence control any migration of the additive.

The migration of any additive within the polymer matrix is found to be affected by factors that include:

a)   the type and the size of the migrant,

b)   temperature, and,

c)   the nature (permeability) of the matrix.

In (c), for example, it is known that filled systems permit less migration (and thus, non-carbon-black filled rubbers have the highest values of migration).

The low-density of plastics is an advantage for their use in general, but, at the same time, the relatively loose packing of their molecular system allows the easy permeation of gases and liquids through them. This can be important in many applications, particularly in packaging.

It is not possible to generalise about performance and level of permeabilities, as quantified in terms of the permeation constant (k), because some plastics have high permeabilities (poor in offering resistance to the passage of fluids or chemicals through them), and some have very low permeabilities [2].

There are well-established qualitative and quantitative analytical approaches in characterising migrants from plastics and rubbers (especially for food packaging/food contacting materials) [3, 4]. A few years ago, it was accepted that the plastics industry could use *migration modelling* in the compliance testing of plastic materials. When a calculation by this model confirms that the level of migration of a compound is below the specific migration limit, that is considered as enough documentation for compliance with legislation, while in the case of non-compliance, the results will certainly need to be checked further and verified experimentally [5].

Multidimensional comprehensive environmental evaluation of packaging materials are also possible, including their environmental impact and life-cycle assessment [6].

This chapter will focus on those used as intentional additives for plastics, that can be classified as toxic substances, extremely toxic substances or severe poisons, or which have been shown to be toxic with long-term (chronic) effects and with special focus on those that possess carcinogenic characteristics.

Table 3.1 presents some information for common intentional additives.

Table. 3.1 Functional additives used with plastics and rubbers and their main functions		
Name	Function	Commonly Used Chemicals
Antistatics (electrostatic-discharge dissipating (ESD))	Dissipation of static charges	Mostly amines, quaternary ammonium compounds, phosphate (organic), and polyethylene glycol esters (PEG)
Colorants	Coloration of plastics, dyes, organic and inorganic pigments	Benzidene (diarylide) (yellow), nickel azo (yellow), benzimidazole, copper phthalocyanine (blue-green) and isoindolinone (yellow-orange and reds), heavy metal oxides and sulfides, heavy metals and titanium dioxide, dihydroindolizine (DHI) and thermochromic antimony
Compatibilisers (adhesion promoters)	Improving the interaction between different phases	SBS in styrenic blends, epoxidised or maleated functionalised polymers in general
Cure Agents and Cure Accelerators	Crosslinking the system	Benzoyl peroxide for plastics, zinc oxide and sulfur for rubbers
Coupling Agents	Improving the bond between the matrix and the reinforcement	Silanes and titanates
Foaming Agents	Improving the formation of cellular structure	
Impact Modifiers	Improving impact properties, melt index, processibility, weatherability.	
Nucleating Agents (optical property modifiers)	Increasing crystallinity	

Name	Function	Commonly Used Chemicals
Table. 3.1 Continued...		
Processing Aids (external and internal lubricants, and so on)	Improving processability	
Plasticisers (flexibilisers)	Increasing flexibility, reducing melt temperature, lowering viscosity	
Preservatives (antimicrobials/biocides)	Controlling and stopping microbiological deterioration	
Processing Aids (polymer processing additives (PPA))	Improving processing behaviour and surface finish, increasing production rate, decreasing viscosities	Lubricants, such as wax or calcium stearate; also antiblocking, release and slip agents
Stabilisers	Retarding or inhibiting decomposition by heat, light (UV), oxidation or mechanical shear	
(a) Antioxidants	Resisting oxidation	Phenolics, amines, phosphates, thioesters
(b) Heat Stabilisers	Resisting thermal degradation	Tin compounds
(c) Light (UV) Stabilisers	Anti-ageing	Some polymers
(d) Flame Retardants	Resisting and burning	Chlorine, phosphorus or metallic salts, MDH, ATH
(e) Oxygen Scavengers (oxygen absorbers, oxygen sorbents, oxygen barrier materials)	Resisting any effect and penetration by oxygen	PVDC, PA or EVOH
Others	E.g., abrasion and surface additives	Low molecular weight PTFE, UHMWPE

MDH: magnesium hydroxide
ATH: aluminium hydroxide
PVDC: polyvinylidine chloride
EVOH: ethylene - vinyl alcohol
PTFE: polytetrafluoroethylene
UHMWPE: ultra-high molecular weight polyethylene

3.2.2 Antistatic (Electrostatic-discharge-dissipating) (ESD) Intentional Additives

All polymers are poor conductors of electricity and can concentrate static charge on their surfaces rather easily, causing a number of problems, e.g., electrical shocks experienced by the consumer (or more critically by employees working at the machines), and, in industrial packaging in some extreme cases, even spark-induced fire and explosions. This, in addition to problems encountered during the processing, transportation, storage and handling of sensitive plastic electronic components and devices, as well as the general problem of dust contamination (dust pick-up) that can affect the appearance and performance of the end product. All these should be minimised and avoided. The catastrophic risks can be controlled or avoided by use of proper antistatic ESD agents or conductive fillers, as well as by use of intrinsically conducting polymers can make surfaces (or the bulk) more conductive electrically, thus helping to dissipate high electrical charge densities and reducing the possibility of any spark or discharge forming. By use of ESD, surface resistivities can be increased up to 50 times, independently of humidity. Antistatic ESD agents are also called surface (property) modifiers. Antistatic agents can be either of internal or external type:

Internal ESD agents are added to a polymer during processing (and so exist in the bulk of the system), while,

External ESD agents are applied to the plastic surface after processing (either by spraying or dipping).

Internal ESD are preferable to external, because the latter are comparably more short lived and can be abraded easily from the polymer surface in use. For high-quality applications, e.g., in printing on the surface, an internal type of application is also preferred: during processing, the internal non-permanent ESD migrates to the surface of the polymer, building up a uniform layer with the hydrophilic end of the ESD agent projecting out of the polymer and its lipophilic end anchored in the polymer. Although internal ESD are more efficient and preferable, they can present a greater potential for toxicity, as considerably higher concentrations are used, the probability of migrations are both taken into account. In any case, for use in food contact applications, ESD must have the required food-contact clearances and must satisfy all national regulations.

In addition, certain contaminants in these chemicals (such as toluene, styrene, and so on, which can off-gas onto the wafer surfaces) can damage sensitive electronic components, particularly wafers, prior to die-attachment during processing or package leads.

ESD are commonly referred to as 'antistats' or 'antistatic surfactants', and are mostly low molecular weight ethoxylated amines, quaternary ammonium compounds, phosphates (organic) and PEG esters, ethoxylated esters, and others. They are usually applied in quite large quantities (2% or more), either by compounding directly with the plastics and fibres in bulk, or by applying then directly to the plastics and fibre surfaces by topical coating, to act as a surfactant. As an example of the latter, the coating of PET polyester films with an amphoteric 'sodio-sulfonato polyester dispersion' has been developed to introduce efficient antistatic properties [5].

There are cases where both improved antifogging and antistatic' low-density polyethylene (LDPE) compositions (with low volatile contents) have been prepared by blending the polymer with an erythritol polymer fatty acid ester [6], and antistatic and fire-retardant properties have been introduced together by adding specific agents into the bulk during processing (with PC as the base resin, perfluoroimide metal salts were used [7].

For the application of any antistatic agent, after considering their rather high levels of use in the system in general, it is usually necessary to obtain certain specific, critical approvals (e.g., from the US Food and Drug Administration (FDA)).

Polymeric antistatic agents can provide longer-term performance with greater safety in polymeric systems, *posing fewer health hazards*, because of their structure (polymeric with high molecular weight, hence their migrations *in the bulk of the polymer* are negligible). However, they are still in the process of development. Polyether block polyamide copolymers are such polymeric antistatic agents and can safely provide permanent antistatic properties to a number of polymers, although, large quantities are usually needed (10-20%), which may negatively affect the mechanical properties.

3.2.3 Colorants

Colorants are intentional additives used for colouring plastic products. They are also called *optical property modifiers*.

Colorants can be subdivided into *dyes*, and dispersed organic, inorganic and special-effect *pigments*. The technique for applying the colour to the product usually involves *precolour* (material that is already compounded to the colour desired), *dry colour* (powdered colorant), *liquid colour* or *colour concentrate* (high loading colorant in a base resin).

Dyes are organic colorants, easily soluble in plastics, either directly in the polymer or in a component of the polymer system-with a high potential for migrating out of the system.

Pigments are distinct particulate materials that remain essentially unchanged during the processing and life-cycle of a plastic product.

There are three types of pigments: organic, inorganic and special-effect pigments [8].

3.2.3.1 Organic Pigments

Organic pigments are not soluble in the resin or in common solvents, and so they are compounded with the polymer by evenly dispersing them. Benzidine (diarylide) (yellow), nickel azo (yellow) (used in cellulosics, PVC, and polyolefins), benzimidazole, copper phthalocyanine (blue-green) and isoindolinone (yellow-orange and reds) (used with high performance or engineering plastics), and so on are some of the organic pigments commonly used.

3.2.3.1 Inorganic Pigments

Some of the inorganic pigments used are based on heavy metals (e.g., barium, cadmium, iron, lead, mercury/chromium oxides, titanium, zinc, complex inorganic pigments as mixtures of two or more metal oxides, and sulfides), or they can be metals themselves (e.g., aluminium, copper, gold), dispersed as powders into the plastic bulk.

Use of inorganic pigments has been criticised, because they can leach from the plastic and easily pose a health hazard and even if such a plastic is discarded into landfill after use, heavy metal ions can migrate out and spoil the groundwater. Incineration of coloured plastic waste is not a solution for these systems, because metal residues remain in the ash. In a number of states therefore, some organic pigments (those containing cadmium, chromium, lead and mercury), or even all of them, are banned from use in plastics for packaging. In any case, where they are used, their concentrations must be well below 100 ppm.

Some of the 'metallic' inorganic pigments are used for their *non-optical effects*. They can provide a complete optical barrier against visible, infrared (IR) and UV radiation.

For protection from nuclear radiation, particularly at short wavelengths, heavy metal inorganic metallic pigments (e.g., lead) are used.

It should be noted that metallic pigments can also provide both electrical and thermal conductivity properties to the plastic material involved, hence they can let the system serve as an antenna and microwave absorber, for a number of specific applications.

Some of the inorganic pigments are *carcinogenic*, e.g., nickel-containing pigments are Group 1 carcinogens, carbon black is a Group 2B carcinogen, and cobalt-containing pigments are labelled as 'possibly carcinogenic to humans' (Group 1A) which means that there is sufficient evidence to regard the substance as causing human cancer, and Group 2B signifies a known carcinogen for animals and possibly a human carcinogen as well, while some are *non-carcinogenic* e.g., trivalent chromium compounds with permissible exposure limit and threshold limit values of 0.5 mg/m³. Titanium dioxide, red iron oxide, and blue cobalt oxide inorganic pigments are also in the safe group, having not been shown to pose any health hazard to humans.

Some supplementary information on the health effects of heavy metals and heavy metal ions is presented in Section 3.3 and in Chapter 10.

3.2.3.3 Special-Effect Pigments

There are pigments with special effects, which can be either be organic or inorganic compounds.

Certain organic compounds like DHI can develop colour in sunlight and lose it in the dark (photochromic material). DHI is commonly used with a range of different polymers, such as: polymethylmethacrylate (PMMA), poly-*n*-butylmethacrylate (P(nBMA)) and polystyrene-polybutadiene (PS-BD) copolymers. Their applications include eye-glasses, light modulators, inks, paints and optical waveguides.

3.2.3.4. Thermochromic Materials

There are some special organic-inorganic compounds that can exhibit reversible colour-changes with temperature (*thermochromic materials*). Certain complex inorganic salts, like $Ag_2Hg_2I_4$ and Cu_2HgI_4, show such reversible changes. In recent years, certain thermochromic compounds containing antimony (green to yellow from 200 °C to room temperature) and arsenic (yellow to red from 116 °C to 295 °C) have also been introduced. Certain plastic systems, such as unsaturated polyester resins, when mixed with Co(II) chloride solutions, are known to gain certain thermochromic properties between 40-70 °C [9].

It is also worth mentioning the special group of *pearlescent pigments,* which are mostly inorganic metals or metallic oxides (e.g., titanium dioxide-mica pigments) [8].

3.2.4 Curing Agents, Cure Accelerators, Crosslinkers (XL)

Curing agents are intentional additives and are used to crosslink resinous systems to improve bulk properties, and ultimately to reach the thermoset state. For under-cure of the system or excess application of curing agents, so that some of the curing agents remain, migration of these chemicals (which are toxic to different extents) may occur.

For plastics and rubbers, and specifically for unsaturated polyesters, peroxides are used as crosslinkers e.g., benzoyl peroxide (BPO) at high temperatures and methyl

ethyl ketone peroxide at room temperature cure for plastics, while for rubber, sulfur is more commonly employed. For rubber, zinc oxide is used (at about 2% or 4 phr) either as a cure accelerator, in most cases, or directly as a crosslinker as well in halogen or carboxylic containing polymers. Zinc oxide is toxic and water-soluble, and is labelled by the EU as 'a dangerous chemical as regards the environment' in category 'N'. Other accelerators used are, in general, metal salts of organic acids namely, cobalt soaps, or tertiary amines e.g., dimethyl amine, applied at 0.05%-0.5% concentrations.

For rubbers, the vulcanisation process is complex, with several different resulting products. Certain rubber cure accelerators can be left completely unreacted in the system (such as thiurams – (tetramethyl thiuram disulfide and tetramethyl thiuram monosulfides, thioazoles, sulfenamides, diphenyl guanidine and dithiocarbamates), and have been shown to lead to a generation of nitrosamines, which are known to be carcinogenic during vulcanisation. Hence, their existence in certain rubber products (such as teats for baby feeding-bottles) is extremely important and levels of nitrosamines in rubber are restricted (<10 µg/kg or as extractable nitrosamines, <1 µg/kg) [10]. In aqueous media there can be additional breakdown products of the curing agents.

The accelerator, zinc dibenzyl dithiocarbamate, used mainly in latex formulations, is permitted under recent German recommendations for consumer goods, both in Category 1 (covering food-contact applications, such as elastomeric seals for food containers,) and in Special Category 2.5 (covering teats, soothers, balloons and so on) at up to 0.5%. German legislation also recognises that not all secondary amines are carcinogenic - those that are not derived from dibenzylamine and dicyclohexylamines, are considered to be 'safe amines'.

Sometimes curing agents, antioxidants and accelerators can interact, creating new toxic chemicals e.g., when guanidine accelerators are used during the vulcanisation of rubber with sulfur with phenylene diamine-based antioxidants, aromatic amines and isothiocyanates can be produced, both of which are suspected carcinogenic agents [11].

3.2.5 Coupling Agents and Compatibilisers

Coupling agents are intentional additives used to enhance the bonding between the matrix and the reinforcement or filler, mostly in polymer composite systems. Fillers can be *reinforcing wetting* (e.g., inorganic glass fibres or flakes), where coupling agents are used effectively to modify the interfacial interaction between the filler or reinforcement and the polymer so that the mechanical properties of the system are improved, or fillers can be used for cost-effectiveness purposes only, in which case they are called *extenders* (where a large volume of plastic can be produced with relatively little actual resin) and these types of coupling agents are not functional chemically. Calcium carbonate, silica and clay are frequently used as extenders.

There are a variety of silanes, titanates and zirconates used as coupling agents for the reinforcing type of filler.

Compatibilisers are intentional additives, incorporated into multi-component, multi-phase polymer systems. They are usually block copolymers, whose segments are soluble in different components of the mixture. Compatibilisers can be reactive (if they form bonds with one of the polymers in the mixture) with reactive groups like acrylic or methacrylic, maleic anhydride, or glycidyl methacrylate), or non-reactive. The main classes of compatibilisers are: (a) modified PE and polypropylene-styrene containing polymers, (b) macromonomers, (c) silane-modified materials.

3.2.6 Foaming (Blowing) Agents

Chemical blowing agents are special intentional additives that decompose easily into gaseous products during the foam processing at temperatures below the processing temperature of the polymer, where a high volume of gas is liberated but trapped within the melt, creating a foamed-cellular structure in a number of different polymer systems (e.g., PVC and (unplasticised PVC), PE, PS, PP, EVA (ethylene vinyl acetate), PET, PC, acrylonitrile-butadiene-styrene (ABS), EPDM (ethylene-propylene-diene-terpolymer), and blends of PVC-nitrile-butadiene rubber, and so on)

Foaming agents are either liquids or solids. In the case of liquid foaming agents, there is in general no decomposition of the type mentioned previously. Instead boiling occurs, the melt foams spontaneously and this structure is captured as the melt freezes or cures finally to a solid structure.

There are a number of options available to select the blowing agent. Eight basic materials are used worldwide:

a)   Azodicarbonamide (ADC), which is the most widely accepted one with a market share of 85% in the EU and is used for foaming most thermoplastic and rubber materials,

b)   4,4-Oxybis benzene sulfonyl hydrazide (OBSH),

c)   *p*-Toluene sulfonyl hydrazide,

d)   5-Phenyltetrazole,

e)   *p*-Toluene sulfonyl semicarbazide,

f)   Di-nitrosopentamethylene tetramine (DNPT),

g)   Sodium bicarbonate, and,

h)   Zinc carbonate.

Usually, the decomposition of any of these is incomplete, although certain catalyst activators can be used to accelerate their decomposition, some residue usually remains in the system (e.g., for ADC, decomposition is 32% by weight, which is mostly nitrogen and carbon monoxide, while the solid residue is mostly urazole and cyanuric acid. One use of ADC is in the production of blown vinyl wall-coverings. As another example, OBSH, which is mostly used in the rubber industry and in the foamed insulation of cables, decomposes with a rather low yield, leaving mostly non-polar oligomers in the residue [12].

Among these agents, DNPT seems to pose the most serious issues of toxicity and also carries the intense smell of the residue, so that although it is the lowest cost blowing agent available, its use is declining.

For polyurethane (PU) systems, water (or traces of humidity) can be used as the foaming agent. Directly injected liquid carbon dioxide is also used as a foamer, as well as a plasticiser. After release of the pressure constrainer of the system, the carbon dioxide boils, causing the polymer to foam [13].

Hydrochlorofluorocarbons (HCFC) are one of the foaming agents used widely in the past, which have now largely been phased out in the developed world, followed by the earlier ban of another ozone-layer damaging chemical, chlorofluorocarbon (CFC), in response to the Montreal Protocol. However, even after the ban of CFC and phasing out of HCFC, some of the PU foam industry is still using an alternative, polyvalent version of HCFC, namely HCFC-141b, which has also been banned in the USA since the start of 2003, and in the EU and Japan since the beginning of 2004.

Recently, there has been a new trend towards the use of a much safer compound as foaming/blowing agent: hydrofluorocarbons [14].

3.2.7 Stabilisers

Stabilisers are intentional additives that help to retard the decomposition of plastics by heat, light (mainly UV), oxidation, and mechanical shear during processing and use. Antioxidants are required for almost all polymers, especially ABS, PE and PS. Heat stabilisers are required during the processing of PVC, because its molecules are very susceptible to thermal degradation.

Hindered amine light stabilisers (HALS) are the main stabiliser type (as a scavenger to inhibit free radical chain propagation), while organo-nickel compounds (as a quencher to prevent initiation of polymer degradation) are used for UV stabilisation.

Stabilisers can be divided into five subcategories:

a) Antioxidants,

b) Heat stabilisers (thermal stabilisers),

c) Photo/light (UV) stabilisers,

d) Flame/fire retardants, smoke suppressants, and,

e) Oxygen scavengers.

3.2.7.1 Antioxidants

Oxidation of plastics and rubber, mostly by heat and/or radiation, results in the breaking of chains and bonds (termed chain scission), and hence a deterioration of mechanical properties as well as changes in the chemical characteristics of the chains. Some polymers, such as PE and PP, are especially susceptible to oxidation, and must be specifically protected.

Antioxidants are usually in the form of *antioxidant packages*, combining two or more antioxidants (one of these, in general, is the *primary antioxidant* while the other is the *secondary antioxidant,* and both are expected to work together synergistically). The primary antioxidant helps to prevent or to terminate oxidative reactions. The secondary antioxidant helps to neutralise reactive materials that can create new cycles of oxidation. Antioxidants are also regarded as 'anti-ageing additives' [15].

Among common antioxidants for plastics, there are phenolics and amines (primary antioxidants), and phosphates and thioesters (secondary antioxidants).

Vulcanised rubber has for many decades, been stabilised by *p*-phenylene diamines (PPD), as antioxidant and antiozonant. However, traditional PPD tend to migrate to the surface and discolour non-carbon black vulcanisates, as they are capable of extraction from the rubber and volatile. In general, all amine antidegradants are soluble in water and need special precautions and environmental labelling.

There are also phenolic antiozonants, based on PPD derivatives commonly used for rubbers.

3.2.7.2 Heat (Thermal) Stabilisers

PVC is one of the lowest-cost commodity resins, used extensively today due to its excellent chemical and mechanical properties. However, the thermal decomposition temperature of PVC is close to its processing temperature, and hence use of proper heat stabilisers during its processing is essential. In addition, PVC-wood composites (plastic lumber) discolour with time during outdoor use, and this can be improved by use of proper thermal stabilisers [16].

Some thermal stabilisers also have an activating influence on the blowing agent during the production of cellular plastics.

Three main families of heat stabilisers are available for PVC:

a) Compounds of lead - basic lead sulfate and lead stearate. These are of relatively low cost, however, it is known that all forms of lead are extremely toxic to humans because of their cumulative effects,

b) Organo-tin compounds - mono- and di-butyltin as well as thioglycolate, have excellent thermal stability and very low toxicities,

c) Cadmium, and complex-salt systems of barium-zinc and calcium-zinc. In these mixed Group II metal complex systems, it is known that, cadmium can cause kidney damage and anaemia, and so the phasing out of cadmium-containing heat stabilisers is underway.

Among these heat stabilisers, those containing tin (such as tin mercaptides, and tin carboxylates or maleates) are considered to be the most efficient, and can be used in a wide variety of applications. Metallic tin is harmless but there are suspicions that organotin-compounds can be toxic to the central nervous system and the liver. However, tin stabilisers have a low capacity for migrating, and hence they are still considered to be safe. Tin stabilisers such as methyl and octyl tins are used in food contact applications. For PVC, it is believed that tin stabilisers act as HCl scavengers (generating tin chloride) as well as an antioxidant. Thio-tin compounds (preferred for rigid pipe extrusions and profiles (for window frames) of PVC) may develop an odd odour due to sulfur.

In addition to these basic heat stabilisers, there are others to consider:

a) *Calcium and zinc systems* (non-lead thermal stabilisers with mixed metals), which offer more environmentally friendly alternatives, specifically for cable-coating applications: for example, the series of calcium-zinc powder stabilisers (self-lubricating as well as non-lubricating) developed for high-temperature cable and automotive wire applications (meeting most North American and European automotive cable insulation requirements), which are non-toxic to humans and can offer comparable properties, though at a higher cost.

b) *Heavy metal stabilisers* (whose fate depends on a number of complex factors - recently some organic-based stabilisers with a pyrimidine-dione system with no heavy metals was also introduced, and they found an immediate use).

3.2.7.3 Photo/Light (UV) Stabilisers (Stabilisers for Photo-Oxidation)

Photo/light stabilisers are anti-ageing additives. It is commonly known that PVC plastic windows can develop yellowish-brownish spots with time, mainly caused by sunlight and other environmental factors (humidity, pollen, iron, rust, soot, and so on). Some polymers, such as polyphenyl acrylate and poly(*p*-methylphenyl acrylate), can be used as photo-stabilisers for the protection of PET. HALS are commonly used photo-stabilisers (as scavengers to inhibit free-radical chain propagation), and in addition to organo-nickel compounds (as quenchers to prevent initiation of polymer degradation) they are used for UV stabilisation. In a blend of HALS with some (partially hindered) oligomeric amines, a synergistic effect in stabilisation is reached, for example, for stabilising PE [17, 18].

3.2.7.4 Flame/Fire Retardants and Smoke Suppressants

Most polymers are composed of organic molecules, and hence are readily flammable. Flame retardants are applied to retard the flame and retard the burning of the plastic system.

It is known that fire evolved from polymers is a very rich source of toxic gases that can cause poisoning and suffocation. In the UK alone, according to data from the Office of the Deputy Prime Minister, there are more than 400 deaths due directly to fire each year, mostly from furnishings and bedding (the initial materials ignited). In the EU each year, there are about 20,000 fires due to the explosion of televisions alone, causing at least 160 deaths and 2,000 serious injuries. New fire-safety and environmental regulations, particularly in the EU, have therefore been established. The EU market for these additives is currently estimated at about $500 million.

Fire retardants can function either by:

a)   Inhibiting ignition,

b)   Suppressing smoke, or

c)   Self-extinguishing.

Flame retardant additives mostly contain bromine, chlorine, (bromine and chlorine being the most effective), phosphorus, sometimes with antimony-based (antimony trioxide) synergists, or other metallic compounds, in addition to hydrates, such as ATH and MDH, and alumina trihydrate, zinc borate, phosphate esters and chloroparaffins. Most of these were developed after the ban on halogen-containing retardants. The ban was put in place because of the toxic nature of halogens and especially their emission in the gas phase when the system is heated, however, the pressure to maintain the ban in Europe has abated significantly. 'Zero halogen' flame-retardants are mainly used for cable applications, where they are known as powerful flame-retardants and smoke suppressants and are specifically used in electrical cable applications of polyolefins. In principle, any halogenated molecule with a high halogen content can be effective, however, halogenated (bromine, chlorine or fluorine) compounds in combination with antimony trioxide are found to be much more effective systems for providing flame retardancy. Each of these fire retardant agents is in general applied at rather high concentrations (55% to 65% by weight), which usually affects the mechanical properties of the material, as well as posing a greater potential health hazard danger where there is any toxicity.

Chlorinated paraffins (mainly CPVC) are widely used in PVC to give greater resistance to ignition and combustion than general-purpose plasticisers. However, the effects of chloroparaffins on health are still a controversial issue and their use as flame-retardants in PVC applications for cables, wall coverings and flooring is declining.

During the last decade, environmentalists have fought strongly to ban the use of brominated fire retardants, and already a number of plastics processors have voluntarily switched to non-halogenated ones (such as phosphate esters, aluminium trihydride and magnesium hydroxide fire retardants). Still, there are two contradictory forces in the fire retardant industry, one is the constant push for stronger fire-safety standards, and the other is the move to eliminate flame-retardants seen as persistent, bioaccumulative or toxic [19].

A recent work shows that an improved, cost efficient and safer new generation of brominated flame-retardants are possible by eliminating the use of antimony trioxide [20].

Furthermore, the human health section of the EU Risk Assessment has concluded that the fire retardant additive TBPBA carries no risks, and the EU Scientific Committee on Health and Environmental Risks (SCHER) has confirmed the EU (Risk Assessment) conclusions that TBPBA presents no human health risk. TBPBA is a brominated flame retardant (used in electrical equipment including computers, televisions, and in printed wiring boards (PWB), and so on [21].

Halogenated polymers, where halogen groups are attached on the main polymer chain with strong bonds, are good candidates for introducing flame retardancy to polymeric systems. Their higher molecular weights provide a number of additional advantages:

a)   Low volatilities,

b)   Less potential for migrating and hence low toxicities,

c)   Much greater ease of handling, and,

d)   Long-term performance.

Most of these properties are very difficult to achieve with lower molecular weight standard compounds.

PP, for example, can be made flame retardant by incorporating a mixture of a sterically hindered alkoxyamine stabiliser and Melamine.

A polycarbonate derivative (tetrabromobisphenol A polycarbonate) and brominated polystyrene are common flame retardants for PBT and PET, and PA, respectively.

PTFE containing halogen (fluorine) atoms, is an effective flame retardant by itself for PC and ABS.

Polymeric flame-retardant systems are introduced for better performance and specifically for better environmental care. They include:

a) *Polymeric brominated flame retardants* (PBFR), which are poorly soluble in water, and limited in migration by their high molecular weights, thus preventing any leaching in the finished product. In any case, their penetration of cell membranes and bioaccumulation in living tissues is very unlikely [22].

b) *Non-halogenated polymeric flame retardants,* such as high molecular weight silicone rubber (used in EVA/PE-based low-energy cables in combination with zinc borate).

Flame retardant chemicals are shifting towards halogen-free products. The fire-safety standards adopted so far by the EU over a wide range of end applications are to ensure the use of flame retardant chemicals that are safer to human health.

As an alternative to harmful fire retardant, nano-clay applications have recently been used very successfully, especially with PU foams that are used in furniture [23]. A draft list of 'Persistent, Bioaccumulative, and Toxic' substances, providing limits under the proposed EU-REACH regulations for 'authorisation' (targeted for control and/or substitution) in the short-term, as well as elimination in the longer term, is also available.

3.2.7.5 Oxygen Scavengers (Oxygen Absorbers, Oxygen Sorbents, Oxygen Barrier Materials)

Oxygen scavengers are used in packaging to help to preserve the contents of the package. Oxygen, if it interacts with the contents of a food package, can affect the taste, colour, odour, appearance and quality of the food. Any residual oxygen that remains after the package is sealed, as well as any oxygen that infiltrates the seals or walls of package, is expected to be absorbed and eliminated by oxygen scavengers. Hence, the shelf life of packaged foods, beverages, pharmaceuticals and other products will be extended.

There are *'passive' oxygen barrier materials,* such as glass, metal, PVDC, PA or EVOH, which can slow down or inhibit oxygen permeating a package. However, on their own these barrier materials are not sufficiently effective because, like butylated hydroxyl anisole - butylated hydroxytoluene (BHA-BHT), sorbates or benzoates, they cannot absorb the residual oxygen that can remain trapped inside the package, or within the product itself, thus leaving the contents vulnerable to spoilage.

Oxygen scavengers are usually *'active' barrier materials* that help to produce packaging systems safe from deterioration by oxygen. Absorption of oxygen in a food container can retard the growth of aerobic bacteria, moulds and other agents of spoilage. Oxygen scavengers can also reduce or even eliminate the use of chemical preservatives like BHA-BHT, sorbates or benzoates.

Scavengers can either be directly compounded with the plastic material, or encased in porous packets (sachets) and placed inside the sealed plastic container. Iron oxide and unsaturated fatty-acid salts are usually applied in sachets, however, incorporation of scavengers into the packaging material leads to more uniform and dependable properties.

Oxygen scavengers can be applied effectively in single-walled monolayer systems such as systems in the form of PET bottles have been shown to be a good choice for beer bottling, because beer is very sensitive to oxygen and spoils quickly when exposed to it.

3.2.8 Impact Modifiers

Impact modifiers help to improve impact properties by toughening the system, as well as improving the melt index, processability and weatherability.

As an example, rigid PVC provides an excellent cost/performance balance with inherent flame retardancy. It is used particularly in building and construction applications such as window frames, doors, fencing and plumbing, but it needs improvement in some mechanical and weatherability properties, which is achieved by use of an acrylic-styrene or a butadiene-based, rubbery *impact modifier*.

Impact modifiers function by dispersing *a damping phase* capable of absorbing energy (to stop craze propagation) into the brittle matrix, and in general, elastomerics are preferred and used.

Acrylic copolymers (i.e., *core-shell impact modifiers* with a shell of PMMA and a core of butyl acrylate elastomer) have been developed mainly for impact modification of PVC for outdoor applications. Butadiene-styrene copolymers are used exclusively for PVC, PC or styrene-acrylonitrile (SAN). Thermoplastic elastomers in the form of styrenic copolymers, e.g., SBS, are used preferably for styrenics and PA. Polyolefins, like EVA, are used for impact modification of technical polymers.

Processing aids are usually based on high molecular weight acrylic copolymers (for PVC). They modify the rheology and processing characteristics of the melt. *Lubricants* are processing aids that function to ease the process and are of two types: either internal lubricants (that influence the viscosity, such as calcium stearates) or external lubricants (such as oxidised polyethylene wax). Lead-stabilised PVC lubricants are a part of the stabiliser system. They are important in PVC foam formulations.

3.2.9 Nucleating Agents

Nucleating agents help to increase crystallinity in crystallisable polymer systems, and thus they also can be considered as 'optical property modifiers'.

3.2.10 Plasticisers (Flexibilisers)

Plasticisers are a special group of low molecular weight, man-made organic compounds which are used to increase flexibility, resilience and processability of polymer systems. However, their use presents technical challenges on issues of migration, evaporation, degradation and heat resistance.

In early applications, various natural oils were used to plasticise pitch for waterproofing ancient boats. Modern plasticisers, however are mostly man-made organic chemicals. They are most obvious in the phenomenon of windshield fogging, which is observed particularly in new cars and is due to the plasticisers in the plastics and the PU cushions used in the interiors.

Migration of plasticisers, in particular, has been a fundamental debate due to their use for years in food packaging, as medical plastics and for special applications such as children's toys, because most of them have been shown to be toxic for humans to a certain degree. There are developments that include studies to reduce (or to stop) plasticiser migration, either by using high molecular weight (polymeric) plasticisers usually polyesters, based on adipic acid or by using so-called green plasticisers (**Table 3.2**), as well as by use of plasticisers that are already fixed covalently on the polymer chain. In this connection, low molecular weight copolymers of PS have been used as plasticisers for PS and styrene-isoprene-styrene (SIS) copolymers successfully. However, for such systems there is a problem of cost [24].

Plasticisation, in principle, is achieved by molecules having both polar and non-polar groups on the same chain. The balance between these different functional groups determines the performance of the plasticiser.

Plasticisers may function either externally or internally in preparation and action, and thus can be of internal or external types.

A rigid polymer may be internally plasticised by chemically modifying the polymer chains with structural groups incorporated through a plasticising comonomer, which is a common way of plasticisation or it can be externally plasticised, simply by blending it with the resin. The latter being the most common because of the costs involved. UPVC (PVC-U) can be externally plasticised by use of certain phthalates.

Table 3.2 Some common plasticisers and green plasticisers	
Chemical name	Abbreviation
Some Phthalate-Based Plasticisers	
Bis(2-ethylhexyl) phthalate	DEHP
Diisononyl phthalate	DINP
Bis(*n*-butyl) phthalate	DnBP, DBP
Butyl benzyl phthalate	BBP, BBzP
Diisodecyl phthalate	DIDP
Di-*n*-octyl phthalate	DOP or DOP
Diethyl phthalate	DEP
Diisobutyl phthalate	DIBP
Dimethoxy ethyl phthalate	DMEP
Dimethyl phthalate	DMPh
Di-*n*-hexyl phthalate	
Some Adipate-Based Plasticisers (*for low temperatures and resistances to UV*)	
Bis(2-ethylhexyl) adipate (also called dioctyl adipate)	DEHA/DOA
Dimethyl adipate	DMAD
Dimethyl hexyl adipate	DMHA
Monomethyl adipate	MMAD
Trimellitates (Trimellitate Esters) (*with low volatilities, used where high temperatures are involved*)	
Trimethyl trimellitate	TMTM
Tri-(2-ethylhexyl) trimellitate	TEHTM-MG
Tri-(*n*-octyl, *n*-decyl) trimellitate	ATM
Tri-(heptyl, nonyl) trimellitate	LTM
n-octyl trimellitate	OTM
Maleates	
Dibutyl maleate	DBM
Diisobutyl maleate	DIBM
Sebacates	
Dibutyl sebacate	DBS
Benzoates	
Epoxidised Vegetable Oils	
Sulfonamides	
N-Ethyl toluene sulfonamide (*ortho* and *para* isomers)	oETSA, pETSA
N-(2-Hydroxypropyl) benzene sulfonamide	HP-BSA
N-(*n*-Butyl) benzene sulfonamide	BBSA-NBBS

Table 3.2 Continued...	
Chemical name	Abbreviation
Phosphate Esters	
Tri-*ortho* cresol phosphate	TOCP
Glycols/Polyethers	
Triethylene glycol dihexanoate	3G6, 3GH
Tetraethylene glycol diheptanoate	4G7
Polymeric Plasticisers	
Polymeric adipates	
Phthalic polyesters	
Caprolactone derived polyesters	
Other Chemicals (*able to function as plasticisers*)	
Nitrobenzene	
Carbon disulfide	
β-Naphthyl salicylate	
Some safer (green) plasticisers (*with better biodegradability and no biochemical effects*)	
Acetylated monoglycerides	
Alkyl citrates	
Acetyl tributyl citrate (compatible with PVC and vinyl chloride copolymers)	ATBC
Acetyl triethyl citrate (with a higher boiling point and lower volatility than TEC)	ATEC
Acetyl trihexyl citrate (compatible with PVC)	ATHC
Acetyl trioctyl citrate	ATOC
Trihexyl *o*-butyryl citrate, butyryl trihexyl citrate (compatible with PVC)	BTHC
Epoxidised soya bean	ESBO
Tributyl citrate	TBC
Triethyl citrate (used for controlled release medicines)	TEC
Trihexyl citrate (compatible with PVC, also)	THC
Trimethyl citrate (compatible with PVC)	TMC

3.2.10.1 External Plasticisers

There are around 25 phthalate compounds, all acting as external plasticisers. Most of them are either persistent organic pollutants (POP) or POP-like compounds.

External plasticisers can be classified either as 'primary' or 'secondary', when considering their effects. A *'primary plasticiser'* improves mechanical properties (e.g., elongation and softness) by flexibilising the systems, however, it cannot usually make these changes alone, because of limited compatibility with the polymer.

'Secondary plasticisers', also known as 'extenders', are used with primary plasticisers to enhance their plasticising performance [25, 26]. For example, esters of fatty acids and monocarboxylic acids in liquid form function as secondary plasticisers for plasticised PVC compounds. There are also several subcatagories of secondary plasticisers, such as 'general purpose', 'high/low temperature', 'non-migratory', 'fast fusing', and 'low viscosity' [27].

3.2.10.2 Plasticisers and Health

Plasticisers should probably be considered major components of plastics rather than ordinary additives.

There are more than 300 different types of plasticisers known, and 50 of them are in commercial use. A classic example is the production of flexible grades of PVC (artificial leather) by use of a series of phthalate plasticisers. In fact, the most commonly used plasticisers are esters, such as adipates, mellitates and phthalates.

(a) *Phthalates and Health*

In Western Europe, about one million tonnes of phthalates are produced each year. About 900,000 tonnes are used to plasticise PVC.

The most common phthalate plasticisers are di-2-ethyl hexyl phthalate (DEHP), di-isodecyl phthalate (DIDP), and di-isononyl phthalate (DINP). Phthalate use has been controversial for years because of its suspected health hazards, especially in the case of DEHP. In both wildlife and laboratory animals, phthalates have been linked to a range of reproductive health effects, with claims that most of them can function as an endocrine disrupting chemical (EDC), and also as cancer-causing agents (specifically in the liver and kidneys). Such negative effects that phthalates are suspected to have on health have been attributed to short ester chains (< C_9). In fact, phthalates are already distributed worldwide in the environment. Some phthalates are even found in deep-sea jellyfish 1,000 meters below the surface of the North Atlantic Ocean. A number of studies have shown that most people are probably contaminated by substantial quantities of these chemicals, and yet for humans, no 'safe level' of exposure to phthalates has been determined. There are also claims about the leaching of certain phthalate plasticisers from biomedical plastics (e.g., intravenous tubes) and hence directly into the patients bloodstream. In one such study, it was shown that about 60% of the DEHP/DOP had migrated to the patient, while almost all of the bis (2-ethylhexyl) adipate (DOA) has been retained in the tube [28].

As a result, criticism has been raised against any use of phthalates at all, as expressed by certain non-governmental organisations, such as Greenpeace and Friends of the Earth [29]. The main phthalates that are under investigation are BBP, dibutyl phthalate (DBP), DEHP, DINP and DIDP.

It is interesting to note that as early as 1971, NASA scientists were warning against the use of PVC in aerospace applications, because of the volatility of phthalates in a vacuum.

Since September 2004 in the EU, it has been illegal to use certain phthalates (DEHP, DINP, DIDP) in toys and personal-care products for children. The EU had set the year 2002 as the key milestone for completing its phthalates risk assessment. Two of these compounds (DINP and DIDP) were shown to cause accelerated dehydrochlorination of PVC at high temperatures [29].

(b) *Concerns Related to DEHP*

DEHP can be metabolised rapidly by most animals into monoethylhexyl phthalate (MEHP) and 2-ethyl-hexanol, and both DEHP and MEHP are believed to be hazardous to humans.

Over the years, various official and unofficial claims concerning DEHP have been published, some of them contradictory. Several examples are given next.

It was once believed that DEHP was safe. In 1999 it was claimed that DEHP was completely safe, even in medical devices and toys [30], on the grounds that most of the research on animals was not relevant to humans because of differences existing between their metabolisms.

In a report in 2000, the International Association of Research on Cancer (IARC), which is attached to the World Health Organisation, reduced the carcinogenic rating of DEHP from 2B to 3 [31-35]. The carcinogenicity ratings of chemical compounds are: 1-carcinogenic effects in humans': 2A possible carcinogenic effects in humans, 2B, 3 and 4 possibly no carcinogenic effects in humans.

Following suspicions that DEHP can act as an EDC, the maximum permitted content in the EU was set at 0.1%, which meant that in practice their use was forbidden. They have been replaced mainly by adipate and citrate esters, in the belief that they present less hazards [36].

Possible risks of toxic effects on the liver from exposure to DEHP were evaluated by University of Lille researchers on using rat hepatocytes in 2005, with the finding that DEHP and MEHP show possible (short-term) cyto/hepatotoxicity [37].

Also in 2005, with effect from 2006, the EU Parliament enforced a complete ban on six phthalates that were being used as plasticisers; DEHP, DBP, BBP, DINP, DIDP and di-*n*-octylphthalate (DOP). The first three are not permitted in any toy or childcare article, while the second three will be banned from toys and childcare products (where children could them put in their mouths). However, the US Consumer Product and Safety Commission has accepted that there is no danger posed by certain phthalate plasticisers like DINP. While the debate in the USA continues, producers in the EU have already began replacing DEHP with its alternatives (since 2000, with phthalates like C_9 and C_{10}) and one of the main producers of DEHP (BASF in Ludwigshafen) shut down its factory in 2005, although it has maintained production in Asia [38]. At the moment, DEHP is the only phthalate that has European Pharmacopoeia approval for its use in the flexible medical devices [29].

Following these moves, there has been a partial Asian ban: Taiwan decided to categorise DEHP (and DOP) as toxic chemicals at the end of 2005, and the China General Administration of Quality Supervision, Inspection and Quarantine has banned the use of DEHA in PVC food packaging [39, 40].

In the same year, in a US government journal, it was claimed for the first time that exposure to phthalates in the womb can affect sexual development in male infants, by suppressing male sexual hormones [41].

The following is a selection of other research findings from the same period: Swedish researchers reported that male workers in PVC plants have a risk of developing seminoma (a form of testicular cancer) that is six times that of the general population. The increased risk appears to be linked to DEHP, which can promote tumours by disrupting the endocrine system. No increase in risk was found among workers manufacturing other types of plastics.

Another study found DEHP at seven times the normal level in a group of Puerto Rican girls, aged 6 months to 2 years, who were showing premature breast development.

A 1999 study in Oslo, Norway, concluded that young children may absorb phthalates from vinyl floor covering-children in homes with such coverings had an 89% greater chance of developing bronchial obstruction and symptoms of asthma than did children living in homes with PVC-free floor coverings.

In 2004, a joint Swedish-Danish research team found that there is a very strong link between allergies developed in children and the phthalate DEHP, however, a study by the US Institute of Medicine, 'Cleaning the Air', concludes that there is 'insufficient evidence of a link' between phthalates, such as DEHP in vinyl flooring, and any health problem, like asthma.

Finally, it is worth mentioning that a recent FDA guidance report has notified hospitals and other health-care providers that a variety of medical devices (such as intravenous tubes) containing DEHP should be withdrawn from certain patient populations, particularly neonatal males, and recommends the use of several replacement materials, although it is also recognised that there is a real, though, minor risk from exposure to DEHP.

(c) *Alternative Safer (Green) Plasticisers?*

Certain alternative plasticisers have been developed, with the aim of creating safe compounds, but these alternatives are significantly more expensive and their technical performance sometimes worse than that obtainable with phthalates. In addition, it is not clear whether these alternatives really offer a reduction in health risk.

One of the safer alternatives suggested as a substitute for phthalates, is soybean oil. Soybean oils are in general more expensive than phthalates, but they also posses the following advantages that phthalates do not:

a)   They confer stability (hence the need for the additional heavy metal stabilisers is eliminated), and,

b)   They do not leach from the plastic, thus diminishing any health risks and indirectly helping to extend the life of the product.

The complete isosorbide family and its esters are also recommended as alternative green plasticisers: their toxicity data show no acute toxicity, no sensitisation, and no mutagenic (Ames test), or oestrogenic effects [42].

A further account of plastics, plasticisers, PVC and connected health issues is presented in Chapter 5.

3.2.11 Preservatives (Antimicrobials, Biocides)

Preservatives (or antimicrobials) are used to slow or prevent microbiological deterioration of plastic products and their contents, by killing micro-organisms such as bacteria, fungi and algae, or limiting their growth. Antimicrobial packaging shows promise as an effective method for the inhibition of certain bacteria in foods [43, 44].

There are a number of anti-microbial additives available for plastics at ambient as well as at high temperatures.

Ohs gives a comprehensive list of 'biocides' for use with plastics [45].

According to OHS, the following properties are desirable for a biocide:

a)   Low toxicity (to higher organisms and to the environment),

b)   Easy to apply, with no negative impact on the properties or appearance of the plastic object,

c)   Storage stability and long-lasting efficacy.

Bacteria, fungi and algae are well known in the plastics industry for causing problems, such as embrittlement, discoloration, loss of light transmittance and the production of odours, because some ingredients (certain plasticisers, lubricants, thickening agents and fillers) can support their growth. Some plastics can provide a very suitable substrate for the proliferation of pathogenic microbes in critical environments (such as hospitals and restaurants). It is known that, high humidity can trigger microbial attack. There are several biocide additives, also known as antimicrobials or biostabilisers, available to prevent or to reduce such problems. Some common plasticisers used with PVC are: dioctyl phthalate, diisooctylphthalate, dibutylphthalate, tricrescyl and triphenyl phosphate and are already inherent biocides and are resistant to microbial attack. There are also specific antimicrobials used in PVC, such as 10-10′-oxybisphenoxarsine (OBPA), *n*-(trichloromethylthio) phthalimide and 2-*n*-octyl-4-isothiazolin-3-one. Recently, many end-users have been switching to new alternative biocides, such as isothiazolinones.

Silver-based biocides are also gaining a share in plastics markets, especially in Japan, where silver ions supported in inorganic matrix/substrates (such as zeolites or alumina-silica), are often used in hospital settings. Silver-nanoparticle-coated PU foams have been shown to be effective antibacterial water filters [46].

There are a number of test methods available to demonstrate the efficacy of biocides and to test directly whether they might present health hazards [44]. The EU's Biocidal Products Directive plays a significant role, both in selecting/recommending and restricting the use of different biocides [47].

3.2.12 Processing Aids (or Polymer Processing Additives, PPA)

Processing aids are basically used to improve processing or surface finish and to increase production rates e.g., linear low-density polyethylene and its blends with LDPE are increasingly used in place of LDPE in packaging applications, because they offer specific advantages and cost reduction. However, they are also more difficult to process and need certain processing conditions. Certain PPA are helping to solve these problems - antiblocking, release and slip agents are also in this group, and there are more than 800 such additives.

Processing aids are used in general to improve polymer processability and handling, and they function usually by reducing the melt fracture to 'improve surface appearances', by reducing gelling and by improving production rates through pressure reduction, e.g., in an extruder. Usually they create a thin layer between the metal and polymer melt, and hence help the polymer melt to slip off more easily. High molecular weight polydimethyl siloxane for example, is used successfully in most cases to create such surface lubricant layers, and acrylic polymers help the processing of PVC by promoting fusion and lubrication.

Lubricants are a group of special chemicals that aid the processing of plastics. Lubricants such as wax or calcium stearate help to reduce the viscosity of the molten plastic and improve forming characteristics. Hence, the action of *internal lubricants* can be considered as similar to that of plasticisers: they reside in solution within the polymer, decreasing the melt viscosity, and when polymer is solidified the lubricant continues to function by increasing the flexibility. External lubricating agents, on the other hand, are compounds that can function as mould-release agents by reaching the surface of the mould by migration.

3.2.13 Compatibilisers (Adhesion Promoters)

In general, when two or more different types of polymer are mixed, they show immiscibility (separation of phases), and exhibit a dispersion of the minor phase (the dispersed phase) in the major phase (the matrix). The cohesion between these different phases is low in most cases, and so the mechanical performances of the blends produced by simple blending are generally low, and need improvement. This is of prime importance in the preparation of 'multi-layered' products (packaging films, bottles, and so on) or 'multi-material' parts, as well as where plastics are reinforced by fillers or by fibres. In such cases, interfacial cohesion between different phases can be improved by the use of special additives, called *adhesion promoters* or *compatibilisers*, that act at inter-phases.

There are a number of polymeric compatibilisers, e.g., block/graft copolymers like tri-block copolymers of SBS used mostly in styrenic blend compatibilisation, and functionalised polymers with certain functional groups (epoxidised or maleated), which can act like a 'surfactant'.

3.2.14 Other Intentional and Unintentional Additives

Abrasion and *surface additives,* such as low molecular-weight PTFE, can be mentioned alongside the other members of the intentional additives group, because they are used to enhance abrasion resistance, to reduce the coefficient of friction and mechanical wear, and to reduce surface contamination, through being blended as a powder with the system. Moving thermoplastic parts, such as gears, benefit from the reduced friction of the self-lubricating properties gained at the surface.

Ultra-high molecular weight polyethylene (UHMWPE) can also be considered as such an intentional additive, because it can be used for scratch-resistant coatings, which can add scuff resistance to moulded parts and contribute to noise reduction. Both PTFE and UHMWPE can be assumed to be safe as long as their sizes and migration capacities are considered.

Amongst unintentional additives, there are the monomers and oligomers that can be left after the polymerisation reaction as remnants, such as styrene, after the production of styrenic polymers and copolymers, butadiene from polybutadiene polymers or copolymers, or acrylonitrile from polyacrylonitrile or nitrile rubber production, the last being relatively more toxic with stringent limits established (1 mg/kg). The danger of such small molecules (relative to the polymer itself) is that they can easily migrate out, and in the case of rubber, where the matrix is flexible, such migrations is more significant.

New chemicals produced through interactions within the system are also unintentional additives (as discussed in Section 3.2.4): e.g., the thermal breakdown product di-isopropylbenzene when the peroxide 1,3-bis-*tert*-butyl peroxy isopropyl benzene is used as curing agent, is an example.

3.3 Health Hazards of Heavy Metals and Heavy Metal Ions

Some additives can contain certain heavy metals or their ions, for different purposes, (e.g., stabilisers used in PVC window profiles and pipes, which are mostly lead-based or barium/cadmium/zinc compounds, in addition to a number of organonickel compounds used for UV stabilisation). These can pose health hazards if they migrate out of the system.

Certain metals and metal ions are well known for their toxicities. Although all living organisms require specific metal ions for physiological processes, when they are present at concentrations above levels of homeostatic regulation, they can become toxic. Metals can exert toxic effects on functional groups in enzymes.

For this reason, heavy metals and metal ions that are used as additives in plastics and rubber (as colorants, stabilisers, plasticisers and so on) should be monitored carefully, and their use as well as the amounts used should be well known and regulated. In addition to their existence in some of the additives used in plastics and rubbers, toxic heavy metals most of which are considered chronic poisons, such as arsenic, lead, mercury, cadmium, nickel, zinc and chromium, are frequently encountered in industrial processing and other manufacturing operations (their main industrial sources include paint, ink, plastic, rubber and plastic film production, leather tanning, wood preserving, battery manufacturing, and so on).

Heavy metals can also accumulate in soil and eventually enter the food chain through their water-soluble compounds, which poses a potentially severe danger.

The following section summarises some basic information on common heavy metals and their toxicities as regards their use in plastics and rubber. More detailed information can be found in Chapter 10.

3.3.1 Some Elements, Common Heavy Metals and Heavy Metal Ions

One can categorise elements in general as 'human elements' 'absorbed elements' and 'others' by considering their interaction with the human body.

(a) *Human elements*

Human elements are those that exist in the blood, bone and tissues. They are the 25 essential elements that the human body requires to function properly: oxygen, carbon and hydrogen, plus smaller amounts of others such as nitrogen, calcium and phosphorus, in addition to sulfur, potassium, sodium, chlorine, magnesium, silicon, iron, and so on.

(b) *Absorbed elements*

Elements like aluminium, barium, cadmium, lead, strontium, and traces of others exist in different forms in plastics additives, and may enter the body through food and water as well as through air breathed in. These elements do not serve any known purpose in the body, but they are still absorbed, and as a result, the average adult body can contain significant amounts of them. Some of these resemble human elements (e.g., strontium resembles calcium closely, and a lot of it is absorbed easily in bones, to the extent that approximately 320 mg can be found in the body of an average person, which is far more than many of the essential elements, while even gold and uranium can exist in quantities of 7 and 0.07 mg, respectively), and are retained and deposited preferentially either in the skeleton (e.g., uranium, binds specifically to the phosphate of the bones) or elsewhere for example in the liver, where liver proteins can trap and deposit some of these heavy metals, like cadmium.

(c) *Others*

Food elements are required on a daily basis because the body cannot retain them, but needs a regular supply.

Cosmic elements, i.e., ones that are naturally abundant in their elemental form in extraterrestrial environments, are mostly hydrogen and helium.

Medical elements are prescribed by doctors to treat diseases or for diagnostic purposes, many of which are regarded as dangerous.

Environmental elements make up the earth's crust and also exist in soil, dissolved in water and the oceans, and in the atmosphere as gas. Environmental pollution and SBS increase the number and concentration of new additional elements.

War elements are associated in various ways with war e.g., antimony (a component of Greek fire), arsenic (a component in mustard gas), beryllium (evolved through nuclear bombs), carbon, (in gun powder), chlorine (the main ingredient of poison gas), iron (swords), manganese (in armour), nitrogen (in explosives), phosphorus (in bombs an nerve gases), plutonium (from classical the atomic bomb), sulfur (in gunpowder), uranium (in atomic bombs), and vanadium (in armour) [48].

Most of the environmental elements that the human body is exposed to normally can be considered as those that exist in the earth crust (oxygen, silicon, aluminium), elements that exist dissolved in the oceans (chlorine, sodium, magnesium, strontium and so on) and those in the atmosphere as gas (nitrogen, oxygen, argon, and radon and so on).

3.3.1.1 Aluminium

Aluminium is a 'non-essential' element for anything living and is abundant in soil, and hence plants absorb a lot of it. Its absorption by the body is prevented by silicon and enhanced by citric acid, which forms a soluble compound. Aluminium, once in the body, is difficult to remove, but is still not regarded as a poison at normal levels (approximately 60 mg in the whole body). Aluminium oxide is used as a flame-retarding additive in plastics.

3.3.1.2 Antimony

Antimony is a 'non-essential' element and has no biological role. The total amount existing in the body is normally about 2 mg (100 mg is the lethal dose).

Among antimony compounds are *antimony hydride* (stibine, a gas) which is the most deadly, followed by *antimony sulfide* (stibinite), which is used in modern times in camouflage paints (because it reflects IR), and the *antimony oxide/lead oxide and carbonate* mixture *Naples yellow*, was used as a pigment and paint as well as a flame retardant in plastics (PVC) for their use in car components, in televisions and so on, because it quenches the fire by reacting chemically with burning materials, although it was also accused of causing cot deaths (see Chapter 10).

High levels of antimony can still be found in some old houses, probably due to the old lead-piping systems as well as layers of lead paints that contain antimony.

3.3.1.3 Arsenic and its Compounds

Arsenic is known as a deadly poison, but it is also an essential (trace) element for some animals (and may possibly be so for humans), with a necessary intake of 0.01 mg per day! It is suspected that arsenic in small doses stimulates the metabolism and boosts the formation of red blood cells. However, it has been claimed that prolonged exposure to fumes of arsenic compounds, mainly *arsenic trioxide,* causes skin and/or lung cancer.

Arsenic is one of the few compounds (besides vinyl chloride) that causes the rare liver cancer: *angiosarcoma.*

Arsenic derivatives were used as pesticides and as paint (by artists: the bright yellow pigment, *royal yellow*, which was favoured by seventeenth-century Dutch painters, is known to oxidise slowly to the deadly compound *arsenic oxide*), and also in wallpapers (during the nineteenth century, most wallpaper was printed with emerald *green* containing *copper arsenate*, which could easily be transformed into the deadly gas *methylarsine*, by a particular mould that developed when the walls became damp, causing arsenic poisoning.

Wood treated with copper arsenate (CCA timber or Tanalith) is used as a structural and outdoor building material.

Arsenic has been called the 'Poison of Kings' and the 'King of Poisons'.

Elemental arsenic and arsenic compounds are classified as toxic, dangerous and category 1 carcinogens in the EU under *Directive 67/548/EEC.*

The IARC recognises arsenic and arsenic compounds as *Group 1*, and the EU lists arsenic trioxide, arsenic pentoxide and arsenate salts as *'Category 1'* carcinogens [49, 50].

3.3.1.4 Barium and its Compounds

Barium is a 'non-essential' element for humans, and has no biological role. Barium compounds are mainly used in glass making, in medical diagnostics, in textiles, and in oil and gas exploration.

3.3.1.5 Bismuth and its Compounds

Bismuth has no biological role in the human body, and poses no environmental threat to living species at normal concentrations.

Certain bismuth compounds are used as catalysts in manufacturing the monomer acrylonitrile and also in the production of bismuthoxychloride, whenever a pearl effect is needed.

3.3.1.6 Boron and Boron Compounds

Boron is an essential element for plants and humans (we absorb about 2 mg daily through fruits and vegetables). The total amount of boron in the body is about 18 mg, 5 g of boric acid can make a person ill, and 20 g or more is potentially fatal.

Sodium borate (or borax) is well known as a skin ointment (along with mercury, lead and sulfur).

Boron compounds are used in making glass and detergents as well as being used in agriculture.

Borosilicate glass fibres are used in reinforced plastics and as insulation in buildings.

Boric acid is used in ceramic glazes for tiles and kitchen equipment.

Boron compounds are also used as food preservatives, and boron compounds such as sodium octaborate are used to fireproof fabrics. Boric acid is also used as an insecticide, in particular for ants and cockroaches in the home. Compounds of borate are used in many other applications (e.g., as fertilisers, in face powders to add a lustre and a silky feel, and so on).

3.3.1.7 Bromine and its Compounds

Bromine exists as the molecule Br_2, with no biological role in humans.

Polybrominated diphenylether is used as a flame retardant for furniture foams, plastic casings and some textiles, but since it can mimic hormones, it is regarded as potentially dangerous.

3.3.1.8 Cadmium and its Compounds

Although there are claims that cadmium can be essential to health to some extent, this has not been proved completely. However, it has been shown that cadmium stimulates the metabolism. It exists in blood and bones, and the total amount in the human body can be 20 mg at age 50, increasing with age.

Cadmium is an accumulative poison and is on the UN Environmental Programmes list of the top 10 hazardous pollutants.

Cadmium mimics the element zinc (which is essential to the body), and hence it is absorbed and tends to accumulate in the kidneys, to the point when they can no longer function properly if the level exceeds 200 ppm. At this point it begins to prevent re-absorption of proteins, glucose and amino acids, damages the filtering system hence leading to kidney failure, and is particularly damaging to the testicles. Cadmium, once absorbed, can remain in the human body for 30 years. Excess exposure to cadmium can lead to weakening of the bones and joints, making movement painful (the itai-itai disease in Japan).

There are also concerns that cadmium causes cancer in humans, although this is not completely proven.

Cadmium sulfide is a common pigment (the bright cadmium yellow), and although it has been banned for some time, it used to be added to paints, artists colours, plastics, rubbers, printing inks and vitreous enamels.

Until recently, cadmium red was the preferred choice of red pigment for containers, toys, and household items, replacing pigments derived from toxic heavy metals such as lead and mercury, but cadmium is now considered totally undesirable and its pigments are being phased out, probably to be replaced with cerium sulfide, which gives a rich red colour, is stable up to 350 °C, and is completely non-toxic.

Cadmium oxide fumes are the most dangerous cadmium compound with inhalation having a number of fatal consequences.

3.3.1.9 Caesium

Caesium has no known biological value, but it can replace potassium, for which reason in excess amounts it can be toxic. There is approximately 0.6 mg of caesium in the body and the average daily intake is 0.03 mg through food and even through tea leaves (3 ppb). Caesium is used in industry as a catalyst promoter and is used to make optical glass. Caesium gel is widely used in Japan for skin rejuvenation, where it probably acts by boosting enzymes.

3.3.1.10 Carbon and its Compounds

Carbon is a cosmic and human element. There is about 16 kg of carbon in the human body, most of it in the tissues. It is the most essential element to life. Carbon itself is non toxic, but some simple compounds (carbon monoxide and cyanide) can be very toxic. Carbon black, which is like soot, is an irritant and can harbour carcinogenic materials.

Carbon powder is used in plastics (and especially in rubber) as reinforcing filler, and its carcinogenic effect is still under debate.

3.3.1.11 Chlorine and its Compounds

Chlorine in its chloride ion form is essential in the human body. There is about 95 g of chlorine in the body, and at least 3 g of it is needed per day. Chlorine and chloride are (relatively) non-toxic. The element itself, as chlorine gas, is very toxic (it is perceptible in air at 3 ppm concentration, and breathing air with 500 ppm of chlorine for five minutes can be fatal).

There is evidence that white blood cells in the human body use this gas to defend against infection.

Chlorine gas was used as an offensive weapon during World War I. Chlorine gas is now used in water purification to make it safe to drink or to swim in. However, this practice is still under discussion because there may be the possibility of chlorine reacting with organics in the water to form ppb traces of organochlorine products. It has been shown that about 2000 organochlorine compounds are also produced in nature, by algae, plankton, trees, and fungi, as well as volcanic eruptions (about 75% of the organochloride compounds existing in nature). It has been found that some microbes and bacteria can dechlorinate and hence decontaminate sites that are heavily contaminated by man-made organochlorines. Most organochlorides (like polychlorinated biphenyls) are regarded as potentially damaging to human health and are persistent in the environment and hence pose a threat.

Organochlorines and all other chloroform-type chemicals have a set limit of 100 ppb in drinking water.

A compound of chloride, hydrochloric acid, is classified as a dangerous chemical above certain concentrations. However, it is also known that HCl is produced in the human stomach normally at low concentrations to help to break down food and to destroy bacteria.

About one-quarter of chlorine produced is used in the production of PVC and in the production of flame-retardants for plastics.

3.3.1.12 Chromium and its Compounds

Chromium is a 'human element'. There is a trace amount (1-12 mg) in the body, which is needed to utilise glucose, and it is also involved in an unknown type of interaction with RNA. Lack of chromium can lead to diabetes, however, high concentrations, especially of chromates, are labelled as 'extremely toxic' (and are suspected as being carcinogenic). Chromium workers usually develop 'chrome ulcer' disease. Compounds containing hexavalent chrome are considered serious human health hazards. OSHA, NIOSH, and CDC determined that calcium chromate, chromium trioxide, lead chromate, strontium chromate and zinc chromate specifically cause (lung) cancer [43].

Chromium and its compounds are used as catalysts in polymer production and as pigments in plastics. Chromium salts are very colourful and the pigment chrome yellow (lead chromate) was once very popular with artists and designers, because of its brilliance.

3.3.1.13 Lead and its Compounds

Lead-based paint with lead carbonate is bioavailable (highly soluble in bodily fluids), while lead chromate has very low bioavailability, with no carcinogenic potential.

Atmospheric lead emissions are mainly from the exhaust gases of cars (1.3 kt in the UK in 1997), from the use of lead tetraethyl additives in petroleum.

Lead is a neurotoxin that induces intellectual dullness, reduced consciousness and in extreme cases, damages the blood structure, gastrointestinal, nervous and reproductive systems and in the longer term, causes coma and death.

The atmospheric exposure limit for lead is 50 µg/m^3.

3.4 Regulatory Bodies for Heavy Metals and Metal Ions

There are certain regulatory bodies, such as the Superfund Amendment and Reauthorisation Act (SARA) and the Resource Conservation and Recovery Act (RCRA) of the USA, the first of which (Title III SARA) regulates the toxic inventory and emissions, while the latter (RCRA) regulates disposal of hazardous waste in general. In addition there is the Clean Air Act (CAA), which regulates the abatement of all materials in the air, and OSHA, which regulates exposure to chemicals in the workplace. The Clean Water Act (CWA) controls the limits of metal concentrations in water.

Metals and metal ions that are regulated by these bodies are presented in **Table 3.3**.

The pigments regulated for use and disposal that are most affected are antimony, arsenic, cadmium, hexavalent and trivalent chromium, lead, and nickel compounds. Lead, arsenic and cadmium-containing pigments are nearly obsolete now as plastics packaging materials. In the heavy metal family, solubility in bodily fluids is the key to toxicity.

3.5 Toxic Chemicals from Degradation, Combustion and Sterilisation of Plastics and Rubbers

Plastics and rubbers can decompose rather easily under various environmental effects (such as heat and UV light), if stabilisers are not used. Decomposition products are

Table 3.3 Leading regulatory bodies for some hazardous metal and metal ions					
Regulated by:	SARA	RCRA	OSHA	CWA	CAA
Aluminium (Al)	no	no	no	yes	no
Antimony (Sb)	yes	yes	no	yes	no
Arsenic (As)	yes	yes	no	yes	yes
Barium (Ba)	yes	yes	no	yes	no
Cadmium (Cd)	yes	yes	yes	yes	yes
Copper (Cu)	yes	no	no	yes	no
Lead (Pb)	yes	yes	yes	yes	yes
Mercury (Hg)	yes	yes	no	yes	yes

mostly toxic volatile organic compounds, and are usually due to the degradation of the polymers involved and, to some extent, the various additives used.

This topic will be discussed at the end of the following chapter, and various possible toxins that can evolve from polymers and can damage human health are considered.

3.6 Effect of Migrant Compounds on Taste and Odour

Although not directly related to human health, a number of additives in plastics and rubber (as intentional, unintentional and interaction chemicals) have objectionable, offensive odours, which affect the quality of life [51]. These additives include some monomers e.g., styrene, in addition to most amines e.g., tertiary amine catalysts used for PU foams, which have strong odours and are also blamed for the fogging of windows in automotive interiors, phenols, mercaptans, peroxides, aldehydes, ketones, alcohols, and some plasticisers and fire retardants. Some organic solvents (used during plastics processing) can also give off strong chemical smells. When these compounds migrate into food from packaging or from any plastic or rubber material in contact with food, even at ppm levels, they can spoil its organoleptic characteristics, and should be eliminated. The reduction or elimination of these chemicals is accomplished by a number of techniques [52]:

(a) *By replacement (of unpleasant smelling additives with less odourous substitutes)*

Some of the tertiary amines used for PU can be replaced by using catalytically active polyols, which can replace up to half of some tertiary amine catalysts and can thus result in reduced odours.

Phenol stabilisers used with PVC are replaced by less odourous, zinc-based heat stabilisers. Octyltin heat stabilisers, with their reduced odour and reduced fogging properties, are used favourably with PVC in automotive applications.

PPA slip-agents of certain reduced odour amides e.g., erucic and oleic acid-based vegetable-derived amides, are used in polyolefin and styrenic food-packaging materials.

(b) *By minimising levels of monomers in plastics*

Since the traces of some remaining monomers can give off unpleasant odours in many plastics and rubbers e.g., monomers of PVC, PS, polyvinyl alcohol and acrylics, they can be eliminated or kept to a minimum by using special inherently monomer-free polymers.

(c) *By using odour absorbers*

Small amounts of synthetic zeolites (based on metal aluminosilicates) can be added to the system to adsorb unwanted organic odour-producing molecules and/or to remove the moisture that contributes to odours (by trapping them in their highly porous crystal structures). This technique is successfully used in extruded polyolefin pipes, injection and extrusion blow-moulded containers, barrier packaging materials, extrusion coatings, and sealant polymers.

(d) *By using antimicrobial agents (to prevent the emission of musty odours by bacteria and fungi)*

Antimicrobials like OBPA, trichlorohydroxy diphenyl ether (Triclosan), as well as *n*-octyl-isothiazolinone, 4,5-di-chloro-isothiazolinone, mercaptopyridine-*n*-oxide (pyrithione), and butyl-benzisothiazolinone are frequently used to reduce odours.

Organometallic compounds of tin and silver are also used in most cases to reduce odours and to retard surface growth, staining and embrittlement.

(e) *By other methods (desorption, rinsing, fragrance use)*

External odour absorbing agents can be used e.g., activated charcoal or high-surface-area silica. This may be done under a vacuum to speed up the process.

Rinsing plastics in special formulated (detergent) solutions (aqueous alkaline solutions with surfactants) can also help in odour removal, in particular in the elimination of vinyl monomers, styrene, acrylates and acrylics, as well as unsaturated hydrocarbons.

The addition of certain pleasant-smelling fragrances are sometimes also used to masks odours. There are commercial scents available specifically for PE, PP, and thermoplastic elastomers, although they may not last for the lifetime of the product.

In the processing of recycled plastics, solvent extraction and degassing can help to reduce odours considerably.

References

1. A. Landuzzi and J. Ghosh, *Journal of Plastic Film & Sheeting*, 2003, **19**, 3, 173.

2. M.W. Kadi, *Asian Journal of Chemistry*, 2005, **17**, 1, 40.

3. B. De Meulenaer and A. Huyghebaert in *Handbook of Food Analysis, Volume 2: Food Science and Technology*, Ed., L. Nollet, Marcel Dekker, New York, NY, USA, 2004, p.138.

4. B. De Meulenaer and A. Huyghebaert in *Handbook of Food Analysis, Volume 2: Food Science and Technology*, Ed., L. Nollet, Marcel Dekker, New York, NY, USA, 2004, p.1297.

5. T. Kanda and M. Fujita, inventors; Mitsubishi Chemical Polyester Film Co, assignee; JP 2004175821A2, 2004.

5. J.H. Petersen, X.T. Trier and B. Fabech, *Food Additives & Contaminants*, 2005, **22**, 10, 938.

6. K. Ri, H. Fujita, H. Kawakami and H. Tajima, inventors; Rengo Co., assignee; JP 2004099721A2, 2004.

6. C.C. Huang and H.W. Ma, *Science of the Total Environment*, 2004, **324**, 1-3, 161.

7. T. Isozaki, inventor; Idemitsu Petrochemical Co., assignee; JP 2004083831A2, 2004.

8. *Coloring of Plastics: Fundamentals,* 2nd Edition, Ed., R.A. Charvat, Wiley-Interscience, Hoboken, NJ, USA, 2003.

9. R.W. Heseltine and J.B. Dawson, inventors; Pilkington Brothers, assignee; US Patent 3,723, 349, 1973.

10. C. Cardinet and H. Niepel, *Revue Generale des Caoutchoucs et Plastiques*, 1994, **729**, 64.

11. H.J. Kreztchmar in *Proceedings of a Conference on Hazards in the European Rubber Industry*, Manchester, UK, 1999, Paper No.6.

12. *Chemical Blowing Agents: Providing Production, Economic and Physical Improvements to a Wide Range of Polymers*, Omnexus Articles, 2003, *www. omnexus.com/resources/articles/article.aspx? id=2&or=p101_11631_101_2*.

13. *Low Environmental Impact Polymers*, Eds., N. Tucker and M. Johnson, Rapra Technology Ltd, Shawbury, Shrewsbury, UK, 2004.

14. V. Enaux, *Urethanes Technology*, 2006, **23**, 2, 22.

15. T.Y. Ching, J. Solis, R. Abbott, W. Diecks and D. Landry, inventors; Chevron Phillips Chemical Company, assignee; WO 2005/026020, 2005.

16. R. Bacaloglu, P. Kleinlauth, P. Frenkel and P.Reed in *Proceedings of the 62nd SPE Annual Technical Conference* - ANTEC 2004, Chicago, IL, 2004, Volume 3, p.3931.

17. C.T. Lee, C.H Wu and M-S. Lin, *Polymer Degradation and Stability,* 2004, **83**, 3, 435.

18. R.A. Jasso, M.L. Duarte and A. Berlanga, *Journal of Applied Polymer Science,* 2004, **92**, 1, 280.

19. *Urethanes Technology*, 2006, **23**, 2, 35.

20. R. Borms, R. Wilmer Roland, M. Peled, N. Konberg, R. Mazori, B.Yaakov, J. Schienert and P. Gerolette in *Fire Retardancy of Polymers: New Applications of Mineral Filler*s, Ed., M. LeBras, The Royal Society of Chemistry, Cambridge, UK, 2005, p.399.

21. *EU Confirms No Classification of TBPBA for Health*, Omnexus News, 12 October 2006. *www.omnexus.com.*

22. L. Zonnenberg, R. Borms and P.G. Eurobrom in *Proceedings of a RAPRA Conference - Addcon World 2004*, Amsterdam, The Netherlands, 2004, p.99.

23. L. White, *Urethanes Technology,* 2006, **23**, 1, 32.

24. K.K. Hansen, C.J Nielsen and S. Hvilsted, *Journal of Applied Polymer Science,* 2005, **95**, 4, 981.

25. D.F. Cadogan, and C.J Howick, in *Encyclopaedia of Chemical Technology,* Volume 19, 4th Edition, Ed., S.R. Kirk and F. Othmer, John Wiley & Sons, Hoboken, NJ, USA, 1996, p.258.

26. D.F Cadogan, and C.J Howick in *Ullmann's Encyclopaedia of Industrial Chemistry, Volume A20*, 5th Edition, John Wiley & Sons, Hobokon, NJ, USA, 1992, p.439.

27. Y. Ito, H. Ogasawara, Y. Ishida, H. Ohtani and S. Tsuge, *Polymer Journal,* 1996, **28**, 12, 1090

28. Q. Wang and B.K. Storm, *Polymer Testing*, 2005, **24**, 3, 290.

29. S.G. Patrick, *Practical Guide to PVC*, Rapra Technology Ltd., Shawbury, Shrewsbury, UK, 2005.

30. S. Toloken, *Plastics News (USA)*, 1999, **11**, 19, 1.

31. *Japan Chemical Week*, 2000, **41**, 2065, 11.

32. *Plastics and Rubber Weekly*, 2000, **1827**, 3.

33. *Handbook of Plasticizers,* Ed., G. Wypch, ChemTec Publishing, Canada, 2004.

34. *Macplas International*, 2000, **5**, 61.

35. *Italian Technology*, 2000, **1**, 27.

36. *Gemini*, 2004-2005, 6.

37. *High Performance Plastics*, 2005, 8.

38. *Plastics and Rubber Asia*, 2005, **20**, 136, 8.

39. *Plastics and Rubber Asia,* 2005, **20**, 138, 20.

40. R. Higgs, *Plastics News (USA)*, 2005, **17**, 19, 8.

41. S. Toloken, *Plastics News (USA)*, 2005, **17**, 14, 3.

42. *Green Plasticizer*s, Kimpeks, *www.kimpeks.com/modules.php?name=News&file =article&sid=18*

43. K. Cooksey, *Food Additives & Contaminants*, 2005, **22**, 10, 980.

44. D. Nichols in *Proceedings of a RAPRA Conference - Addcon World 2004,* Amsterdam, The Netherlands, 2004, p.51.

45. D. Ohs in *Plastics Additives Handbook*, 5th Edition, Ed., H. Zwiefel, Hanser, Munich, Germany, 2000, p.647.

46. D.A. Shafer, *Hazardous Materials Characterisation*, Wiley-Interscience, Hoboken, NJ, USA, 2006.

47. K. Aschberger in *Proceedings of Stick!, 3rd European Congress on Adhesive and Sealant Raw Materials*, Vincent Verlag, 2003, Nuremberg, Germany, p.183.

48. J. Emsley, *Nature's Building Blocks: An A to Z Guide to the Elements*, Oxford University Press, Oxford, UK, 2001.

49. *Chemical Industry Archives*, Environmental Working Group, www.chemicalindustryarchives.org/dirtysecrets/arsenic/1.asp.

50. A. Meharg, *Venomous Earth*: *How Arsenic Caused the World's Worst Mass Poisoning*, Macmillan, Basingstoke, UK, 2005.

51. M.J.M. Brown, J.L. Licker and M.R. Zbuchalski, inventors; No assignee; US 20050129812A1, 2005.

52. G. Gordon, *How to Keep Plastics Odors Under Control,* Omnexus Trend Reports, 2005, *www.amnexus.com/resources/articles/article.aspx?id=7850.*

Bibliography

1. FDA document 21 CFR 179. 45, sub-part C, USFDA gives a list of approved packaging materials including approved additives and specification of several packaging materials for use with gamma or x-ray sterilisation. www.cfsan.fda.gov/~dms/opa-pmna.html, 2005.

2. ASTM F1640-03, *Standard Guide for Packaging Materials for Foods to be Irradiated,* 2003.

3. M. Schiller, W. Fischer, A. Egger, P. Wilhelm and A. Gupper *Gummi, Fasern, Kunststoffe,* 2004, **57**, 5, 302.

4. J.M. Batt in *Proceedings of PVC 2002: Towards a Sustainable Future*, Brighton, UK, 2002, p.496.

5. E.S. Stevens, *Green Plastics*, Princeton University Press, Princeton, NJ, USA, 2002.

6. P.A. Toensmeier, *Plastics Engineering,* 2002, **60**, 11, 7.

7. L. Van Wabeeke, *International Polymer Science and Technology,* 2002, **29**, 2, 1.

8. E.W. Flick, *Plastics Additives: an Industrial Guide*, 3rd Edition, William Andrew, Norwich, NJ, USA, 2002.

9. *Conductive Polymers and Plastics*, Ed., L. Rupprecht, Plastics Design Library, Norwich, NY, USA, 1999.

10. G. Scott, *Polymers and the Environment*, Royal Society of Chemistry, Cambridge, UK, 1999.

11. A.J. O'Driscoll, E. Kramer and A. Gupta in *Proceedings of International Conference on Additives for Polyolefins*, Houston, TX, USA, 1998, p.293

12. R. Kohlmanand, J. Joo and A.J. Epstein in *Physical Properties of Polymers Handbook*, Ed., J.E. Mark, AIP Press, Woodbury, NY, USA, 1996.

13. A.K. Bakshi, *Bulletin of Material Science,* 1995, **18**, 5, 469.

Web Sites

European Council for Plasticizers and Intermediates: *www.ecpi.org*

Phthalate Esters Panel of the American Chemistry Council: *www. phthalates, org*

4 Toxicity of Rubber and Plastics Due to their Non-Additive Ingredients

4.1 General Outline

The main ingredients of rubber and plastics are polymers, which may contain some remnants of monomers left from polymerisation, or which may depolymerise to produce monomers or oligomers of their own, or which may degrade to produce toxic chemicals (if suitable stabilisers and antioxidants are not incorporated in appropriate amounts, or if the systems have been in use for a long time and have aged). As a result of these inherent factors, some toxic substances may already exist in a polymeric systems, before any consideration is given to the possible main toxic effects of all the additives used intentional or unintentionally (discussed briefly in the previous chapter). Remnants of monomers (as well as all other toxic chemicals, including additives that can migrate from rubber and plastics in contact with food) are, by definition, considered food additives, because 'food additives' are those that are added to foods both intentionally and unintentionally. Hence, it is of utmost importance to examine the possible existence of these chemicals (monomers, oligomers etc.) and their health effects, separately from the effects of other additives.

Additional information is also provided for some of these chemicals in Chapters 10 and 11.

4.2 Polymers, Monomers, Oligomers

There are two important classes of polymers, as outlined in the previous chapter, if their thermal response is considered: thermopolymers (leading to thermoplastics), and thermosets. Composites and foamed (expanded) structures can be processed from either of these.

Basic aspects of both of the thermopolymer and thermoset types, as well as some of their combinations, will be discussed briefly in this section, focusing on the possible health hazards of their monomers and of any other possible remnants from the reactions involved.

For *all* possible hazardous effects of processed plastic and rubber items, including possible effects due to additives (as outlined in Chapter 3) and due to monomers and other remnants (as outlined in this chapter), this and the previous chapter should be considered together.

4.2.1 Thermopolymers/Thermoplastics

Thermoplastic polymer systems are such that when they are heated, the weak intermolecular forces are easily overcome, so that at a certain temperature the systems become soft and flexible, and, at higher temperatures, even a viscous melt. When cooled down, the system solidifies back to its hard form. Hence the process is reversible, showing the system's heat sensitivity. Examples of thermoplastics are polyethylene (PE), polystyrene (PS), polyvinyl chloride (PVC), Nylons, cellulose acetate, acetal, polycarbonate (PC), polymethyl methacrylate (PMMA) and polypropylene (PP).

A number of thermoplastics with comparable mechanical properties are considered to be engineering materials.

4.2.1.1 Polyolefins (PE, PP) - the Polythene Family

Polyolefin is a generic name for the polythene family of PE and PP. The burning of these plastics can generate several volatiles, including formaldehyde and acetaldehyde, both of which are suspected carcinogens.

Polyethylene (PE), with the structural formula $[\sim CH_2CH_2\sim]_n$, also known as polythene, is a product of the monomer ethylene $[C_2H_4]$. Low molecular weight (MW) PE (below 6000) is a fluid or wax-like paraffin. Above a MW of 6000, PE is a partially crystalline, lightweight thermoplastic resin with exceptionally high resistance to chemicals, low water absorption and good insulating properties. At 115 °C, PE changes from a solid to a melt, and above this, extensive oxidative degradation occurs if proper antioxidants are not incorporated with the polymer.

There are three main grades of PE:

- Low-density polyethylene (LDPE, or high pressure PE, which were invented in the 1930s), used mainly for the manufacture of plastic bags.

- High-density polyethylene (HDPE, or low pressure/Ziegler PE, invented in the 1950s) with greater rigidity, heat resistance and chemical stability, used mainly for milk jugs, cleaning supply bottles, and trash bags.

- Linear-low-density polyethylenes (LLDPE, from 1977), have much higher stiffness than LDPE, and have improved flexibility/toughness at low temperatures.

PE has very low compatibility with plasticisers, and indeed does not need plasticisers, but it usually contains UV and heat stabilisers.

For possible migrants in HDPE, there are traces of complex organometallic catalysts (which can be reduced to a minimum by repeated washing), traces of stabilisers [1-5] as well as solvents that are left in the system. However, animal tests showed that no toxic effects are involved in general [6, 7]. It is recommended that benzophenone-based UV absorbents and antioxidants are not used in HDPE designed for packaging foods that contain fat, especially corn oil [3-5].

US Food and Drugs Association (FDA) (1998) approved the use of HDPE in contact with food, subject to the provisions of 21 CFR part 177.

Polyolefin floor coverings (of PP and PE), power cables with PE coverings and HDPE pipes and wall-covering materials, halogen-free LLDPE and crosslinked PE (XLPE) are all suggested as potential alternatives to PVC by Greenpeace [6].

Degradation of PE can occur by heat and/or by UV, generating the gases carbon monoxide, carbon dioxide, methane, acetone and methanol [4]. PE is susceptible to hydroperoxidation during degradation. Commercial PE is not biodegradable.

Amongst copolymers of PE, there are polyallomers (ethylene-propylene copolymers), copolymers with cyclo-olefins and with vinylic monomers [with vinyl acetate: ethylene-vinyl acetate, (EVA), with methacrylic acid, MA, and with vinyl alcohol: ethylene - vinyl alcohol, (EVOH)], and chlorinated PE (CPE). CPE, although it exhibits exceptional UV and chemical resistance, gives rise to a high amount of hydrogen chloride gas evolution if combusted.

Crosslinked polyethylene (XLPE), which is PE structurally altered (mainly to improve general thermo-mechanical stabilities) by introducing crosslinks (via radiation or by use of peroxides and silanes), hence with the proper degree of crosslinking XLPE can behave like a thermoset. Crosslinking can be created either during or after moulding. Crosslinking has been shown to reduce the possibility of any migration from the bulk of the material.

Polyethylene foams [expanded polyethylene (EPE), invented in 1941] can be prepared both from HDPE and LDPE, which are usually in thermoset structures (these can be non-XLPE, mostly XLPE or expandable bead structures). EPE is mainly used for insulation and seals. Different azobis compounds are used as blowing agents (yielding nitrogen gas), and different peroxides or silanes, as well as irradiation techniques are used for crosslinking.

Polypropylene (PP) and its copolymer with ethylene are chemically stable and do not need any plasticiser, hence providing and retaining the flexibility without need of any softening agents. PP is preferentially used as a packaging material, as well as in small machine parts, in car components, and directly as fibres.

PP packaging films are usually printed on their surfaces (e.g., in confectionery and snack food products packaging). In this case however, the transfer of ink components can occur from the outer printed surface of the film onto the inner food contact surfaces. If there are plasticisers like dihexyl phthalate (DHP) [or bis-*n*-butyl phthalate (DBP), or di-2-ethyl hexyl phthalate (DEHP)] in the ink, up to 6% of the DHP can migrate to the food immediately (or 6.7 mg/kg for 180 days storage of a chocolate/confectionery product), the proportion increasing with storage time.

A new generation antifog additive has been introduced recently to control the formation of water droplets in food packaging of biaxially oriented PP (BOPP) and cast PP, which offers control of droplet formation in packaging for pre-cut salads and other fresh produce, as well as in hot food packaged before it cools down, such as fresh bread and meat cooked in a store.

4.2.1.2 Styrenics (Styrenic Polymers and Copolymers)

The term 'styrenic' describes the family of major plastic products that use styrene as their key building-block: PS, expanded polystyrene (EPS), acrylonitrile-butadiene-styrene copolymer (ABS), styrene-acrylonitrile copolymer (SAN), styrene-butadiene rubber (SBR) and unsaturated polyester resin (UP). Among these, UP is the only thermoset and will be considered separately in the thermosets section.

- *Polystyrene*

PS is produced from the monomer styrene, and is usually used 'neat', without adding plasticisers. Hence, the health concerns over PS should focus on the remnants of the monomer, styrene.

Styrene (monomer) is an aromatic hydrocarbon with the chemical formula C_8H_8, also known as vinylbenzene, ethenylbenzene, cinnamene, or phenylethylene. Styrene with its high vapour pressure can evaporate easily, and has a sweet smell. It is classified as a Group 2B (possibly carcinogenic to humans) and a human endocrine disruptor (ECD) agent by the US Environmental Protection Agency (EPA) and by the International Agency for Research on Cancer (IARC). Acute health effects of styrene (from short-term exposure) are irritation of the skin, eyes, the upper respiratory tract, and gastrointestinal disorders. Chronic (long-term) exposure affects the central nervous system with symptoms of depression, headache, fatigue and weakness, and can affect kidney function. Studies have shown that exposure to the styrene monomer is associated with menstrual disturbances and neurotoxic symptoms for females [7-12]. The styrene monomer can have general effects that are neurotoxic (by attacking the central and peripheral nervous systems), haematological (reducing platelet and haemoglobin values), cytogenetic (causing chromosomal and lymphatic abnormalities), and carcinogenic. Although the evidence is inconclusive, it is suggested that there most probably is a relation between styrene exposure and a risk of leukaemia. Styrene residues are usually identified in human adipose tissues. Accumulation of styrene in the lipid-rich tissues of the brain, spinal cord, and peripheral nerves, correlate well with the acute or chronic functional impairment of the nervous system. It is therefore recommended that the maximum limit for exposure does not exceed 100 ppm (determined as a ceiling concentration for any 15 minute period). An action level of 25 ppm, one-half of the time weighted average-limit concentration, has been identified to provide adequate protection against the probability of over exposure due to daily variations in environmental levels [9].

Very low levels of styrene exist naturally in a variety of foods (such as fruits, vegetables, nuts, beverages, and meats).

Traces of styrene can affect the quality of food products above certain migration levels. It has been shown that usually over half of the residual styrene in the food contact packaging material migrates within 24 hours, which is the normal shelf life of many food products. The metabolites of styrene are mandelic acid, a known mutagen, and styrene oxide, a known carcinogen.

PS is available in a range of grades varying in impact strength (from brittle to tough).

Non-pigmented PS grade has crystal clarity and a low cost. It is used mainly for the production of vending and yoghurt cups, disposable syringes and casings for ballpoint pens.

PS packages are generally not suitable for use in microwave ovens, especially for products that contain vitamin A (betacarotene), because at microwave temperatures vitamin A decomposes into products like toluene, which can dissolve PS quickly.

PS foam (EPS), which is used as a thermal insulator and packaging material, may contain remnants of styrene. A classical study reported that no styrene monomer is detectable in the air following the installation of rigid PS foam on the inside of rooms used for residential purposes [10], hence concluding there is no danger to health in the use of rigid PS foam indoors. This is one, old, unique example, from which it is not possible to generalise. Remnants of styrene monomer can be a more critical contaminant in other types of EPS foam packages, including foam cups and packages that come into direct contact with food. Since styrene is soluble in oil and in ethanol, if the cup material contains free styrene then ingestion is more likely when EPS foam cups are used to drink beer, wine, or perhaps even coffee with cream.

• *Some Copolymers of PS*

In addition to its use in PS production, styrene is also used to produce a number of copolymers of styrene, such as: ABS, styrene-butadiene-styrene (SBS), styrene-isoprene-styrene (SIS), styrene-ethylene/butylene-styrene, and SAN, and also used at large as the monomeric crosslinking agent in unsaturated polyesters. Hence, the same styrene problem may exist in these copolymers, as well as in unsaturated polyester thermoset systems (discussed shortly in the thermoset plastics section of this chapter).

In the case of SAN and ABS copolymer, the problem of monomer remnants can be more critical, because in addition to remnants of styrene, there may also be remnants of acrylonitrile monomer in the system. These are found to release from their combustion mostly the toxic gases carbon monoxide and hydrogen cyanide [11].

4.2.1.3 PVC and its Copolymers

PVC is one of the most commonly used amorphous thermoplastics. It is available in two forms: plasticised, used mainly in the wire coatings, floor tiles, toy balls, gloves, car seats, rain-wear and in vinyl gramophone-records, and unplasticised, used in pipes, gutters, window frames and wall claddings.

Cellular PVC developed during World War II in Germany, is largely used in structural applications.

PVC is also known as a medical polymer. Pure PVC products are believed to be harmless, and are biologically inert, if prepared and processed properly, like all other polymers. There is no evidence reported so far of the bio-accumulation of PVC.

Pure PVC is thermally and photochemically unstable and has a tendency to lose hydrogen chloride when heated, hence a stabiliser (based on tin, lead, or other heavy metals) is commonly used. Pure PVC is also brittle and needs a plasticiser to soften and flexibilise the system.

Health and safety concerns over PVC are mostly due to the possible leaching out of the residual monomer in addition to certain additives and plasticisers used. In this context, there has been a growing concern over PVC toys (pacifiers and teething rings in particular), that these additives can be released into a child's mouth when the toy is chewed. Old vinyl intravenous (IV) bags, as used in hospitals, have also been shown to leach some DEHP plasticiser. It has been shown that there is also a possibility of the residual monomer and other additives migrating into drinking-water from PVC bottles [13], or from PVC water pipes [14, 15].

PVC is a product of the vinyl chloride monomer (VCM, with the chemical formula $CH_2=CHCl$), which is a well-known carcinogenic chemical and a toxicant.

In addition, most of the plasticisers used for PVC are suspected toxic chemicals.

Hence, PVC products should be checked for levels of both VCM and plasticisers, although it is now possible for the residual monomer to be removed completely from polymer matrices, and in most cases residual vinyl chloride monomer levels in PVC worldwide are well below the maximum acceptable levels, typically from undetectable to a few parts per billion (ppb). PVC resins should almost always be supplied nearly free of any residual or unreacted monomers.

It is claimed that PVC, when burned, produces dioxins, which are known to be deadly poisons and powerful carcinogens [16]. In a recent accident in Seveso, Italy, when four villages were contaminated with dioxins, many people had to be evacuated and the area had to be decontaminated [17]. There is also an example of dioxins being found in cow's milk in the Lickebaert area, Netherlands, which was linked to the incineration of PVC waste in a nearby municipal waste incinerator [18].

In addition, the gas density of PVC smoke is high and it releases mainly corrosive and toxic hydrogen chloride gas. A study carried out to examine the possibility of VCM formation during routine PVC thermal welding revealed that it produced atmospheric concentrations of VCM, as well as benzene, formaldehyde and acetaldehyde, that are well below accepted occupational exposure limits [19]. However, overheating of PVC during processing (e.g., through an overheated extruder) can cause acute upper and lower respiratory irritations due to the toxic hydrogen chloride and carbon monoxide gas emissions [20].

The monomer, VCM, is synthesised from ethylene in the petrochemical industry, through following steps: (1) chlorination of ethylene to 1,2-dichloroethane (1,2-dichloroethane), (2) oxychlorination of ethylene to 1,2-dichloroethane, and (3) cracking of 1,2-dichloroethane to VCM and to hydrogen chloride. Throughout these three steps, about 4 kilograms of by-products are created per ton of 1,2-dichloroethane, containing a number of persistent toxic chemicals, including several of the following so-called 'dirty dozen' persistent organic pollutants (POP): dioxins, furans, polychlorobiphenols (PCB), and hexachlorobenzene (HCB). Any of these, in addition to the VCM residue, can cause environmental problems (mainly during production and processing) as well as health problems (during and after use, if they stay in the bulk of the system as remnants). VCM can stay in the polymer in trace amounts after its production, and can also stay in the (solid) system in small proportions even after processing of PVC into the final shaped products.

In the late 1960s, the carcinogenicity of VCM to humans (specifically in causing liver angiosarcoma) was recognised for the first time from the PVC plant near Louisville, USA, and was also confirmed later by the IARC. In 1964, VCM was still being widely used in beauty parlours as an aerosol hair-spray propellant but although the dangers were realised, the use of them continued for several more years. In 1972, it was officially reported that VCM causes liver cancer in laboratory animals after one year of exposure to levels as low as 250 ppm, and in 1974, Occupational Safety and Health Administration (OSHA) proposed a maximum exposure level for VCM of no detectable concentration, as measured to an accuracy of at least 1 ppm (this was implemented on 1 January 1975).

It was during 1974 that a US FDA ban on VCM use in drugs and cosmetics, followed by a similar ban by the EPA for pesticides, and a ban by the Consumer Protection Safety Commission came into effect. By the end of 1995, 175 cases of angiosarcoma due to VCM had been diagnosed worldwide [21, 22].

The occupational limits for VCM are currently set at 1 ppm averaged over an 8 hour period and 5 ppm averaged over any period not exceeding 15 minutes, with an annual maximum exposure limit of 3 ppm.

The 1997 European Pharmacopoeia sets a maximum of 1 ppm VCM residue in pure PVC.

There are further examples of the possible health effects of plasticisers and additives used in PVC:

- It was shown years ago by Swedish researchers [23] that male workers in PVC plants have six times the probability of developing seminoma (a form of testicular cancer) than do the general population, and this was linked to the ECD agent, DEHP.

- Another study [24] described DEHP present at seven times the normal level in some developmental Puerto Rican girls, aged between six months and two years, who showed premature breast development.

- A 1999 study [25, 26] in Oslo, Norway, concluded that young children may absorb phthalates from vinyl floor coverings, and that these children have almost double the chance of developing bronchial obstruction and symptoms of asthma than do children living in homes with PVC free floor coverings.

As discussed briefly in Chapter 3, various additives used with PVC may also cause health problems, e.g., metal-based stabilisers, which are hard to control in incinerators, and PCB, which are mainly used in PVC for electric insulator formulations and which can cause mammalian infertility.

Waste PVC materials, at the end of their useful life, can continue to cause problems. They are likely to be powerful sources of POP if their waste is incinerated, and dioxins can be produced easily by unregulated burning. The environmental group Greenpeace has advocated the global phase-out of PVC with the claim that dioxins are produced as by-products during the manufacture of PVC, as well as from the incineration of waste PVC in domestic garbage.

It should be noted that the Stockholm Convention obliges all parties involved in dioxin emissions to take all necessary measures to eliminate or to minimise those emissions, and already European Industry has improved production processes in all cases.

The last method for disposing of plastics waste, namely dumping, cannot solve the PVC waste problem satisfactorily, because phthalates and other additives can continue to contaminate groundwater and air, from landfills. Open-air dumps of PVC waste are the most susceptible to leaching (about one-third of the phthalates leaks out under aerobic conditions).

For many years, the effects of PVC on health and on the environment has been debated, sometimes accurately but also opportunistically and with speculation.

It is interesting to observe that although there is as yet no country in the world where PVC has been banned, in certain countries strong anti-PVC lobbies do exist. Paradoxically, PVC is heavily used in the health sector, even for blood bags and tubing for dialysis equipment, and although the details of this debate are beyond the scope of this book, by considering the fact that the production and consumption of PVC increases every year it is not difficult to predict that PVC will continue to be one of the basic plastics for the years to come.

In fact, it can be shown that when proper procedures are applied to the production and use of PVC, products made from PVC will be completely safe without any detrimental effects on health and the environment [27].

Greenpeace has created a page on its website where suggestions for 'PVC Alternatives' are given for those seeking options other than the use of vinyl products [28]. In general, the simplest solution may be to turn to another conventional and a more sustainable plastic. Replacing PVC with appropriate non-chlorinated polymers can virtually eliminate the chances of forming POP, and an even more visionary approach is to use plant origin resins. In fact, there are many alternatives possible for most PVC products, and there are even studies to eliminate the migration of additives from PVC bulk (e.g., by incorporating, usually by grafting, the plasticiser onto the polymer chain, or by use of special high molecular-weight plasticising polymers; such as EVA copolymer, EVA-*co*-terpolymer, ethylene-acrylate-*co*-terpolymer and nitrile rubbers, although the economy of their use is not yet favourable). However, the available evidence indicates that PVC in its applications has no more effect on the environment than its alternatives, and that the possible adverse human health and environmental effects of using PVC are no greater than those of other materials [29]. However, additional studies are certainly very much required for the clarification of some of the issues surrounding the use of PVC.

As a final note on PVC, it is worth mentioning Vinyl 2010, which is an EU voluntary commitment study on the PVC industry initiated in 2001 for the following 10 years, including mid-term revisions of targets in 2005 and definition of new objectives in 2010. The plan includes for full replacement of lead stabilisers by 2015, in addition to the replacement of cadmium stabilisers by March 2001 [27, 30]. Also in Vinyl 2010, the following values are put forward for maximum permissible VCM concentrations in the final PVC products:

For suspension-type PVC, maximum VCM = 5 g/ton of PVC (for general purpose) or 1 g/ton of PVC (for food and medical applications).

For emulsion-type E-PVC, maximum VCM = 1 g/ton of E-PVC.

In the same study, the year 2002 was put forward for the key milestone in completing the phthalates risk assessment [30].

4.2.1.4 Acrylics: PMMA, Polyacrylonitrile (PAN) and Polyacrylamide (PAA)

- *Polymethyl Methacrylate (PMMA)*

PMMA, or Plexiglas as it is commonly called, is made from the monomer methyl methacrylate (MMA). PMMA can usually contain some MMA as a remnant after polymerisation, and MMA can also be evolved from the thermal degradation of the polymer. MMA is a toxic chemical, its health effects including asthma, dermatitis, eye irritation (and possible corneal ulceration), headaches and neurological problems [31-34]. Exposure to very high levels of MMA (>1,000 ppm), which is normally highly improbable indoors, can result in serious neurochemical and behavioural changes, reduction in bodyweight, as well as degenerative and necrotic changes in the liver, kidney, brain, spleen, and bone marrow. Relatively low concentrations of MMA can cause changes in liver enzyme activities. The data concerning MMA's ability to cause cardiovascular effects are, however, inconsistent. Potential risks from MMA for humans mainly arise if there is repeated exposure. The absorption and hydrolysis of MMA to methacrylic acid and its subsequent metabolism via physiological pathways results in low-level systemic toxicity by any route of exposure.

PMMA has exceptional optical clarity and durability for outdoor exposure, with optimum resistance to alkalis, detergents, oils and dilute acids (but not to most solvents), and it is used mostly for glazing as a shatterproof replacement for glass, for lighting, for curtain-wall panels as a sealant and for decorative features, and in acrylic 'latex' paints (containing PMMA suspended in water). There are also PMMA compounds specially developed for production of complex medical components.

- *Polyacrylonitrile (PAN)*

PAN, which is also known as Orlon in its fibre form, is made from the monomer acrylonitrile. Acrylonitrile, with the formula (CH_2=CH-CN), is a toxic, acrid smelling, extremely flammable liquid. In contact with skin or the eyes it produces severe irritation. Inhalation or prolonged skin contact at high concentrations can produce confusion, unconsciousness, and even death. Acrylonitrile is also known as cyanoethylene or vinyl cyanide, and can be found in PAN as a remnant. There are suspicions that acrylonitrile can cause brain cancer [35, 36], and it is classified as a possible human carcinogen (IARC Group 2B).

When burned, PAN releases fumes of hydrogen cyanide and oxides of nitrogen, both of which are highly toxic to humans.

PAN is used in packaging (industrial, medical, and food) and as fibre (Orlon), mainly because of its chemical and UV resistance, as well as for the retention of flavour and aroma and excellent durability under sterilisation procedures especially in packaging. PAN fibres have wool-like properties and are chemical precursors of high-quality carbon fibres. Almost all forms of PAN are copolymers with acrylonitrile as the main component. As copolymers of acrylonitrile, there is SAN to mention which will be discussed next.

- *Styrene-Acrylonitrile Copolymer (SAN)*

SAN is a transparent thermoplastic polymer material with amorphous structure made from the monomers styrene and acrylonitrile. It offers high heat and chemical resistances. The acrylonitrile units of SAN increases the temperature resistance and stiffness with respect to polystyrene. The chemical resistance is also improved through molecular immobility. SAN is quite amorphous and transparent. SAN has very low water absorption and is not attacked by water easily. SAN is preferably used in applications that are in contact with water, which can resist to boiling water as well. It has a particularly wide range of application in electricals/electronics, as well as in reflectors, refrigerator doors and battery cases. However, its brittleness and flow characteristics are a major disadvantage, and either proper elastomer particles (such as, SBS/ABS) or proper thermoplastics (i.e., PVC/PC) are usually blended with it. It is also used as a carrier resin for additives such as pigments to be incorporated in other polymers

- *Polyacrylamide (PAA)*

PAA is an acrylate polymer formed from the monomer acrylamide, which is a neurotoxin that causes peripheral neuropathy. Non-polymerised remnants of acrylamide in PAA pose some potential risks. It is claimed that, people unintentionally absorb around 25 µg of acrylamide daily from the environment, which may account for a significant number of cancer cases. Although there is no solid proof so far for a connection between acrylamide intake and cancer, it remains under investigation, and acrylamide can still be considered as a probable human carcinogen. On a health hazard spectrum, acrylamide registers 2 (3 = a very high hazard to health; 2 = a medium hazard, and 1 = harmful to health; arsenic scores highly at 2.3, while one of the lowest scores is ammonia at 1.0) [37].

PAA is highly water-absorbent and easily forms a soft gel (mainly used in gel electrophoresis, in soft contact lenses and in cosmetic facial surgery).

4.2.1.5 Polycarbonate (PC)

The building block for PC is bisphenol-A (BPA). PC is a tough, durable, shatter- and heat-resistant, transparent plastic. Audio-CD (the first in 1982), followed by CD-ROM and DVD, are all made of PC. In addition to its use in optical data-storage systems, PC provides transparent roofing (e.g., for greenhouses, and the dome of the Sydney Olympic stadium), impact-resistant protective glazing and sheets, as well as structural parts in building and construction. Its application can also range from baby bottles to protective tooth-sealants. PC is favoured for food storage containers because of its clarity and toughness. Being lightweight and highly resistant to breakage, PC has replaced most silica-glass applications, expense remaining its only disadvantage. PC may be seen in single-use as well as refillable plastic water bottles.

PC resins, and BPA as the leachate, are regarded as safe with no substantial health risk to humans [38-41], although there are several studies claiming that BPA can promote certain chromosomal defects (from the production of eggs with chromosomal abnormalities, called *aneuploidy*, which is the leading cause of miscarriages and several forms of mental retardation, including Down's Syndrome). There have also been several animal studies suggesting that BPA can mimic the hormone oestrogen, and hence may affect reproduction, for example by altering the size of the prostate gland and shifting the onset of sexual maturity, although BPA is known to be some 2000 times less potent than oestrogen.

A recent report from the Harvard Centre for Risk Analysis (HCRA) concludes that there is no scientific proof found to substantiate concerns that exposure to BPA can cause reproductive toxicity, and that BPA does not exhibit key characteristics typical of oestrogenic agents [42]. HCRA findings plus decisions made by regulatory officials in Europe and the US argue that exposure to BPA should not pose a health risk.

However, a study conducted up to December 2004 showed that, of the 115 studies on BPA, 94 had found a significant link between low-dose exposure and developmental or reproductive harm [43, 44]. In the same study the authors ask the EPA to conduct a new risk-assessment of the chemical. In addition, another recent report covering the literature on the health effects of BPA leachate at low doses [41] has found a sound correlation, and similarly concludes that a risk assessment is necessary. There is a report from the American Plastics Council indicating that BPA has 1,2-dichloroethane effects on human health, and the California Assembly's Environmental Safety and Toxic Materials Committee has approved a bill to ban the use of BPA in products intended for use by children under three. The HCRA stated that inconsistent results indicate that evidence of 1,2-dichloroethane effects in laboratory animals would not likely be replicated in humans [42].

In any case, BPA can exhibit other toxic effects only at very high exposures and realistically, such high exposures are not possible under normal conditions.

The EU's expert body, the Scientific Committee on Food (SCF), confirmed that PC food-contact applications are safe and meet EU requirements.

However, there is one point of agreement between all these studies and conclusions, which is that sodium hypochlorite bleach and other alkali cleaners should not be used for cleaning at all because they can catalyse the release of BPA from PC.

In addition, some dental composite resins are based on BPA and methacrylates (e.g., BPA-GMA), and there are also initiators (mainly benzoyl peroxide), activators (tertiary aromatic amines) and inhibitors (hydroquinone), as well as certain UV-sensitisers added.

4.2.1.6 Polytetrafluoroethylene (PTFE, Teflon)

Polytetrafluoroethylene (PTFE), also known by its generic DuPont tradename of Teflon, is a linear polymer that has a molecular structure similar to PE, but with all the hydrogen-atoms replaced by fluorine atoms: $(CF_2\text{-}CF_2)_n$. Since fluorine-atoms prefer contact with fluorine-atoms, while repelling all others, this repulsion reduces the friction coefficient of the system and produces a non-adhesive surface (with self-lubricating properties). The strong C-C and C-F bonds make PTFE chemically inert, with excellent chemical resistance.

PTFE is used as a heat-resistant material (for heat-shields on spacecraft), as a non-stick coating (on cooking utensils), and as a lubricant (in bearings, particularly in aggressive environments), as well as for insulating tapes.

PTFE is essentially a non-toxic and non-flammable inert plastic, and can be used over a broad range of temperatures (-200 °C to +260 °C) at which it keeps its properties. It is somewhat stable up to 300 °C, but cooking and bakeware coated with Teflon may still be harmful for human health if the temperature exceeds 260 °C. It has been shown that pyrolysis particles of PTFE are more toxic in most cases than the toxic gases produced, although the fumes at temperatures well over 260 °C can be lethal for pets like birds, which are known to be very sensitive to poisonous gases. Fumes of PTFE, when it is overheated and the decomposition products are inhaled, can cause influenza-like 'fume fever' syndrome (Teflon flu), with coughing, vomiting, and permanent damage to the upper airway.

Thermal decomposition products of PTFE (up to 500 °C) are mainly the monomer, tetrafluoroethylene and some perfluoro compounds (such as, perfluoroisobutylene). Between 500 °C to 800 °C, the decomposition product is mainly carbonyl fluoride, which can hydrolyse easily (to produce toxic and corrosive hydrofluoric acid and carbon dioxide).

The EPA, in 2005, in a draft report, pronounced that perfluoro-octanoic acid (PFOA, also known as C_8), the chemical used to make PTFE, is a 'probable carcinogen' [45].

Meanwhile, DuPont (US) settled for $300 million in a series of 2004 lawsuits based on groundwater pollution from this chemical, and agreed to eliminate releases of C_8 from plants by 2015 [46].

According to the FDA, minute quantities of PFOA can be detected in cookware under extremely harsh conditions that in reality do not compare at all with the conditions consumers use. Proper curing is very important and there should be no measurable amount of PFOA left on a PTFE coated pan, provided that it has been properly cured [47].

On the other hand, implanted PTFE in animals can cause local sarcoma, mainly due to the foreign body reaction.

IARC suggested that there is not sufficient data available to assess PTFE's carcinogenic effect for both animals and for humans.

4.2.1.7 Saturated Polyesters - Polyethylene Terephthalate (PET-S) and Polybutylene Terephthalate (PBT-S)

Saturated polyesters are linear polyesters with high crystallinities, with excellent toughness, strength, abrasion and chemical resistance. There are two common members of this group: saturated polyethylene terephthalate (PET-S) and polybutylene terephthalate (PBT-S), which are composed of large molecules arranged in a linear chain structure that are formed through ester linkages.

Depending on whether there is a linear chain or network structure in the final form involved, either saturated (PET-S/PBT-S), or unsaturated (PET-U/(PBT-U) types are considered. PET-S and PBT-S are thermoplastics (thermoplastic polyesters), while PET-U is a thermoset.

PET-S is formed from the alcohol (ethylene glycol) and the acid (terephthalic acid, or its derivative dimethyl terephthalate, DMT), and has been available for many years as a fibre (e.g., Terylene) and is also used in beverage bottle production (high-impact resistant containers), used as a food packaging material and for carbonated beverage bottles. It has also been used as microwave food trays which can be safely heated in the microwave or in a conventional oven at 180 °C for 30 minutes, although there is considerable concern that additives and remnants may migrate from tray into food during heating, if there is any, and in boil-in-the-bag pouches. PET-S has been used in gears, bearings, housings, in medicine for plastic vessels and for implantation. When melt blown (bottle), it provides a very good barrier for both flavours and hydrocarbons (i.e., fats).

PBT-S on the other hand, is a more attractive moulding material, but is less economical in use.

Some Possible Migrants from PET-S

A number of different migrants from commercial (amber) PET-S bottles have been reported in one study, such as, some intermediates and residual monomers, as well as certain fatty acids and plasticisers [48].

While cyclic oligomers of PET-S (up to 7.0 g/kg) are found in a number of microwaveable PET-S food packages (i.e., those for French fries, popcorn, fish sticks, waffles, and pizza) [49]).

PET-S usually contains detectable amounts of acetaldehyde, which can migrate into the liquid contacting media, usually carbonated beverages, but which does not usually change the taste [50, 51]. Acetaldehyde forms in PET-S rather easily at high temperatures. When acetaldehyde is produced during the processing of PET-S, some of it can remain dissolved in the bottle material and afterwards can diffuse into the product stored inside it, affecting the aroma, which can be very important in the case of PET-S bottled water, because of the new aroma that can develop even at extremely low (10-20) ppb concentrations. This problem is solved by use of small amounts of isophthalic acid as the comonomer during the production of PET-S, which lowers the melting temperature and lets the processing occur at much safer lower temperatures, appreciably preventing degradation.

In one study, about 10% of the acetaldehyde existing in a PET-S bottle is found to migrate out within four days at 40 °C [52-54].

Migration of remnants of catalyst (antimony ion from antimony trioxide (Sb_2O_3) and monomers (ethylene glycol/DMT) into water are not observed, with the exception of migration of some ethylene glycol and antimony ions into food simulants [55-57].

When the Swiss Federal Office of Public Health compared the amount of antimony in waters bottled both in PET and in glass, the concentrations in PET bottles were somewhat higher, but still well below the allowed maximal concentrations (1% of the 'tolerable daily intake' determined by the WHO), and it is concluded that the health risk involved is negligible. A recent study found similar low concentrations for antimony in the water bottled in PET [58].

There is also the most recent WHO risk assessment for antimony in drinking water available [59].

Beyond these facts, it should be also mentioned that, the recycling of waste PET-S is gaining specific attention and there are studies even showing that safe food grade PET can be processed from such resources [58, 59].

As for direct carcinogenicity effect of PET-S films - it is found that they can induce malignant tumours at the site of implantation, however, later, the results of this study have been considered as 'inadequate' [60, 61].

US FDA (1998) approved the use of PET-S in contact with food in accordance with the conditions prescribed in 21 CFR part 177.1630.

4.2.1.8 Other Polymers [Polyamide (PA), Polyurethane, (PU)]

- *Polyamides (Nylons)*

Polyamides can occur both naturally (i.e., proteins, such as wool and silk), or can be made synthetically, like Nylon, Kevlar and sodium polyaspartate. There are several different types of thermoplastic polyamides (PA) such as synthetic polymers (Nylons, e.g., PA6, PA66, PA11), and as a family, they have unique characteristics of high strength, stiffness and toughness, so that they are considered to be engineering plastics, with typical applications including gears, bearings, housings for power tools, slide rollers, and under-the-hood noise reducing/acoustic insulation applications. Another major application of PA is in fibres, which are extraordinarily strong.

In addition, PA6 is a well-established plastic in barrier packaging, providing a barrier to oxygen and aromas as well as to microbials, and is used in food and medical packaging applications. New generation PA6 films with silicate nanoparticles with improved barrier properties are introduced to compete with the conventional barrier EVOH copolymer films.

- *Polyurethanes (PU)*

PU are compounds formed by reacting the polyol component with an isocyanate compound, typically toluene diisocyanate (TDI); methylene diisocyanate (MDI) or hexamethylene diisocyanate. Polyols are relatively non-toxic (i.e., polyether type polyols are found to be safe, because they are low in oral toxicity with almost no irritation effect to the eyes and skin), however, isocyanates are highly toxic and the product can have a significant toxicity if remnants of isocyanate are in it, which manifests itself mainly as a respiratory (as well as a dermal) hazard. Exposure to the vapour of isocyanates directly may cause irritation for the eyes, respiratory tract and skin. Such an irritation may be too severe to produce bronchitis and pulmonary oedema. As health hazards of isocyanates are considered, one immediately remembers one of the worst industrial disasters of the 20th century, that occurred in Bhopal, India, because of the toxic cloud of methyl isocyanate was released accidentally from the Union Carbide pesticide factory in December 1984. An estimated 3,000 people died immediately with a final of some 20,000, most suffocating from the cloud's toxic chemicals, and some 50,000 were injured, most were residents living near the plant.

If PU resins with such isocyanate impurities are allowed to remain in contact with the skin, they may produce redness, swelling, and blistering of the skin, and respiratory sensitisations (an allergic, asthmatic type reaction) may occur, with rather with low dose tolerances [62].

Acrylated urethanes used in dental composites and special sealants in construction and in printing sector are shown as potential allergens [63, 64].

Catalysts used for PU foams are tertiary amines and organometallic compounds, particularly organotin compounds, such as dibutyltin esters. Tertiary amines are strongly basic and usually have high vapour pressures, causing irritation of skin or eyes as well as respiratory system by its vapour.

Although organotin compounds are less of an irritant, any skin contact should still be avoided.

In addition to remnants of monomers and catalysts, PU foams may also have remnants of several other components used during their preparation as well, such as cell-stabilisers, blowing agents, flammability retarding agents and accelerators.

Within the problems that the foam industry face, there are 'the residual chemicals left in the system' and 'flammability and smoke evolution on combustion'.

Any unreacted remnant monomer(s), blowing agent(s), polyols, catalyst(s) and/or fire retardants left and/or trapped in the system inside of the internal cells can be released slowly with time by themselves or as the cells are broken down by use, with the possibility of causing some health problems in both cases. The most widely used blowing agents are azodicarbonamide and azobisformamide, which give off mainly nitrogen, carbon monoxide and ammonia gases during their reaction.

Foams can contain residual polyols as well as isocyanate (TDI and MDI) and hydrocarbon blowing agents. Organic isocyanates are strong respiratory irritants and can initiate asthma-like symptoms for over-exposed persons and chemical sensitisation. Threshold limit values for vapour exposure to cyanates are given as either 0.005 ppm (parts per million) as an 8 hour (for long-term exposure) ceiling limit or 0.02 ppm as a short-term exposure ceiling limit.

Fire hazards of foams, especially that of PU foams, usually present a considerable critical issue. In addition to the PU involved in the main foam structure, many flame retardants (i.e., halogen containing ones specifically) are added to the system. In addition to the foam itself, many flame retardants (i.e., halogen containing ones), contribute to smoke appreciably, and smoke evolution can become a bigger danger than the fire itself. Due to the chemical composition and flammability properties, PU foams, including the flame retarded ones, always generate toxic smoke when involved in the fire, causing narcosis due to the carbon monoxide and cyanic acid (12 ppm) evolved, along with some degree of sensory and pulmonary irritations. Flexible foams (both flammable and non-flammable types) and rigid foams (usually non-flammable) generate smoke with basically average toxic potencies. Toxic gas emissions are generally proportional to the degree of their involvement in fires. Rigid foam, when burned in the flaming mode, produce smoke which is somewhat more toxic. When flexible PU foam cushions are used to assess fire hazards by considering flammability parameters such as ignitability and mass burning/heat release rates, in addition to the toxic potencies involved, it is realised that the overall hazard is relatively insensitive to smoke toxicity as long as the toxic potency values are not too critical [65, 66].

PU is available in three forms: rigid foam (used mainly for insulation in construction), flexible foam (used as cushion material for upholstery) and as an elastomer (used mainly in solid tyres and shock absorbers). All are characterised by their high strength and good chemical and abrasion resistances.

During recent years, applications of plastic cellular materials are mostly in structural and in insulation areas (wall, ceiling and pipe). In insulation of both new and old buildings (retrofit insulation), fibreboard is the most used product for sheathing insulation, and foil-faced isocyanurate foam. In cavity wall insulation, mineral wool, PU, urea-formaldehyde (UF), and fibre glass are widely used. Especially in modular homes, cellular plastics are preferred because of their light weight and insulation capacity. PS and PU foams are used widely for pipe insulation. Cellular rubber and cellular PVC are mainly preferred for small pipe insulation. Urethane-modified and glass fibre reinforced polyisocyanurate foam boards are widely used as insulation materials for buildings. Modified polyisocyanurate foams, with their light weights, high thermal insulations, and strengths, are used on the outer walls of high rise buildings (curtain walls) and for insulation of metal siding (in Japan for the latter, for fire-proof outdoor walls). Extruded PS foam boardstock is used in residential sheathing and in roofing. Both PS and PU foams are preferred for roof insulations.

4.2.2 Thermosets and some Thermoset Composites

Thermosetting plastics (thermosets, in short), unlike thermoplastic materials, cannot be softened by application of heat, and they are completely amorphous and rigid. Thermoset refers to the final form of materials that are cured and hardened through the application of energy (either heat, through a chemical reaction, or irradiation). Curing transforms the system into a thermoset plastic or rubber by the crosslinking process during which, chains are interlinked by strong chemical covalent bonds (crosslinking of the chains), so that the material cannot be softened again by the application of heat (irreversible process). If excess heat is applied, thermal degradation and charring occurs. Common examples of thermosets are amino plastics, phenolics, polyurethanes, epoxies, and unsaturated polyesters, PET-U and vulcanised rubbers.

4.2.2.1 Amino and Phenolic Thermoset Polymers, and their Composites

There are two basic types of amino plastics - urea formaldehyde (UF) and melamine formaldehyde (MF). Both are hard, rigid materials with good abrasion resistance and mechanical characteristics for their continuous use at moderate temperatures (up to 100 °C). UF is relatively inexpensive but can absorb moisture at the expense of poor dimensional stability, and is generally used for electrical switches, plugs, utensil handles and trays. UF foams were mainly applied in construction for cavity-wall type of insulations (foam insulation) during the 1950s.

MF plastics have lower water absorption and better durability, and they are preferentially used for tableware, laminated worktops and electrical fittings. Due to high contents of nitrogen in melamine, MF are non-flammable and can provide fire retardant properties. It releases nitrogen and formaldehyde when burned (or charred) [67]. MF foams were mainly used for fire resistant, soundproofing applications.

Phenol-formaldehyde (PF, Bakelite) is one of the oldest synthetic materials available. It is strong, hard and brittle with good creep resistance and excellent electrical properties, but it is susceptible to attack by alkalis and oxidising agents. Typical applications are domestic electrical fittings, handles, fan blades, smoothing iron and pump parts.

Most amino and phenolic thermosets are used in the production of composites, mainly prepared with wood (wood-plastic composites, WPC) [68]. Particleboard, or the chipboard is a kind of WPC, which is produced from wood particles, such as wood chips or saw dusts, through their glueing together with synthetic resins or other suitable binders, by pressing and extruding afterwards. In most particleboards, the resin used is formaldehyde-based.

There are formaldehyde free resins available, but they are rather expensive and not used too often.

Beginning in the 1980s, there were health concerns about the possibility of toxic formaldehyde vapour emissions from the items made of UF, MF and PF, including their foamed structures, and their use was later discontinued or decreased. In fact, formaldehyde is still one of the most common indoor air pollutants. Formaldehyde in the cured thermoset is believed to be due to that left unreacted, in addition, it may be also be due to a de-methylolation reaction and/or cleavage of methylene-ether bridges. Formaldehyde is a metabolite occurring normally in the human body at small concentrations and it is converted to formic acid by enzymic oxidation. In any case, at concentrations in air, above 0.1 mg/kg, formaldehyde begins to irritate the eyes and the mucous membranes, resulting in headache and difficulty in breathing.

Formaldehyde is classified as a human carcinogen by the US EPA (probable carcinogen), EU (a likely human carcinogen) and IARC (known carcinogen).

Phenolics may have the hazards due to both phenol and formaldehyde. Free phenol is known to have a high skin-absorption potential. Phenol is an irritant (eyes, mucous membranes and skin), and it is a toxic for nervous system as well as to liver and kidneys.

Thermoset composite systems of phenol and formaldehyde, some of which can be substituted for a number of structural applications can also be considered as engineering plastics and they have been in use for a very long time. In recent applications, improved urea - melamine resins have been used as matrices more and more for composite systems, because of their non-melting, high thermal and chemical resistances, hardnesses, mechanical-dimensional stabilities and low flammabilities.

4.2.2.2 Unsaturated Polyesters [Unsaturated Polyester Resins, UP, PET-U]

Unsaturated polyesters are mainly used as the matrix for glass fibre reinforcement, (in glass fibre reinforced thermoset systems). These are most commonly used in the manufacture of small boats, chemical containers, tanks and repair kits for cars.

Unsaturated polyester resins are mainly made by condensing a dibasic acid (1,2-propanediol) with an anhydride (maleic or phthalic anhydrides), by forming ester linkages between the dibasic acid (or their anhydrides) and glycols. Then a reactive monomer (mostly styrene or vinyl toluene, MMA or diallyl phthalate) is used to crosslink the system when needed. 'Unsaturated' denotes the uncompleted chemical activity (double bond) in the original structure, which are used for crosslinking afterwards. In this context, an excess of styrene as the crosslinker (10 to 50 %) is usually added to have it ready in the system, as well as to reduce the viscosity. There are also certain accelerators used (such as, cobalt naphthenate or tertiary amines like dimethyl aniline) to facilitate the cure at ambient temperatures. In addition, there may be pigments, fillers, various inhibitors, accelerators, stabilisers and flame retardants, added to the system. Polymerisation is activated whenever a catalyst (i.e., benzoyl or methyl-ethyl-ketone peroxide) is added.

Reinforced thermoset composites of UP are used extensively because of their non-melting, high thermal and chemical resistances, hardness, mechanical properties, dimensional stabilities and low flammabilities.

Any unpolymerised oligo-polyester left in the composite system has been shown to be responsible for sensitisation and allergic contact dermatitis in use [69, 70].

Health hazards (mostly allergens) observed with UP composites are usually either due to non-crosslinked UP, mainly due to remnants as mentioned previously (such as, styrene, cobalt naphthanate, phthalates or tricresyl phosphate, benzoyl peroxide and other catalysts) or to the under-cured resin (i.e., UP automobile repair putty). Benzoyl peroxide is known to be a strong skin irritant [70].

Inhalation hazards include exposure to toxic chemicals, curing agents, solvents, dust and fumes, as well as thermal degradation products, that are released into the air during formulation of resins-during repair operations-during curing or after curing. In this context, it is known that styrene monomer and other solvents used during production and left afterwards in the system are associated with a number of menstrual disturbances to different degrees, and can lead to chronic illness with secondary amenorrhea, nulliparity, with decreased blood clots (with smoking) and hypermenorrhea (with age) [71]. It is also known that, styrene can cause conjunctivitis and bronchial asthma.

Diethylene glycol maleate that exists in some of the polyester resin cement compositions is known to be an allergen.

4.2.2.3 Epoxy Resins (Epoxides)

Epoxy resins, although introduced over fifty years ago, are still one of the most versatile materials with applications ranging from small domestic appliances to large scale constructions. Epoxides can be a low viscosity liquid or a high melting point solid, depending on the degree of its crosslinking. They have good toughness, small shrinkages during curing, better weatherabilities with low water absorptions, in addition to the outstandingly good electrical properties, heat and chemical resistances, and adhesion. Epoxy resins are a bit more expensive than other equivalent thermosets (e.g., unsaturated polyesters). They, in the form of fibre-reinforced structures, are used preferentially in the aircraft industry.

The epoxy resin based thermosets are either the di-glycidyl ether of bisphenol A (DGEBA), tetra glycidyl-4-4´-methylene dianiline (TGMDA), triglycidyl-*p*-amino-phenol (TGPAP), or the epoxies derived from novolacs. The common curing agents (that polymerise and crosslink) are either a series of aromatic amines (i.e., *p*-phenylenediamine), or aliphatic amines (such as triethylene tetramine), and acid anhydrides (i.e., methyltetrahydrophthalic anhydride).

Within the different types of epoxies, are found 'epoxy diacrylates' or 'vinyl ester' resins, used to produce specific corrosion and chemical resistant composite systems. Vinyl ester resins are produced by either reacting epoxy resins of glycidyl derivatives with methacrylic acid, or from BPA and glycidyl methacrylates, where an active monomer (usually styrene) as crosslinker, hardener (usually organic peroxides), accelerators (cobalt) are added to the system. In the thermoset epoxy systems, there are also the 'mould releasers', which can be either **'internal'** such as, lecithin, or stearates of zinc and calcium, certain organic phosphates that are mixed in the resin, or, **'external'** - such as, fluorocarbons, silicone oil, and certain waxes, that are directly laid on the mould.

Epoxy resins are commonly used for durable and inert coatings, in laminates and composites, and it is used also as an adhesive.

Since properly cured epoxies have relatively high molecular weights and are crosslinked, their vapour pressures are low, hence the potential for their respiratory exposures are normally very low, except when the resin mixture is applied by spraying, or when curing temperatures are high. The common epoxy molecule is a reaction product of epichlorohydrin (ECH) and BPA. Although both (epichlorohydrin and bisphenol A) are suspected ECD agents, generally no appreciable levels of ECH indoors is monitored normally (as mentioned previously), and BPA exhibits toxic effects only at very high exposures which, realistically, is impossible indoors. Hence, properly cured epoxies can be considered to be safe indoors, as far as the related respiratory health hazards are concerned. The potential risks for the dermal (contact) exposure of cured epoxies are, however, more probable than the respiratory exposures. Trace amounts of residual ECH (typically in the range of <1 to 10 ppm, by weight) that may be left in the system, which can cause probable sensitisation, along with the remnants of the hardener, both of which, can cause allergic reactions. Within these curing agents, there are mostly aromatic amines (commonly either 4,4′methylene-dianiline, or 4,4′-sulfonyldianiline), and these compounds with their very low vapour pressures are not expected to present any airborne hazard (unless they are sprayed in a mixture or cured at high temperatures). Within other types of curing agents there are certain aliphatic and cycloaliphatic amines, poly-aminoamides, amides, and anhydrides. In fact, all amine epoxy resin hardeners are shown to present a certain danger to the health of workers, again primarily through skin contact and to a much smaller extent by inhalation [72].

Some low molecular weight chemicals, such as, the conventional epoxy resin hardener methyl tetrahydrophthalic anhydride that can stay as a remnant in the system, can cause 'immunologic contact urticaria'.

Most of the FR epoxy thermoset composites can substitute satisfactorily for a number of structural applications and enable one to support loads. Epoxy thermoset composites are used largely in the construction industry, in other industries, in the production of corrosion proof equipment, in transport industries, as well as for consumer products (i.e., boats, car bodies, pipes, bath tubs, usually to produce large sizes products). They are also used as alternatives to metals in a number of applications. In a recent application, they are used to replace metallic missile parts such as housings for the control actuation systems, warhead, guidance electronics and the wings. This replacement is reported as decreasing the weight of the missile by 50% in addition to increasing the missile efficiency.

Fire retardant epoxy thermoset composites are preferred in most cases because of their non-melting, high thermal and chemical resistances, hardnesses, mechanical and dimensional stabilities and low flammabilities. Glass fibres, in general, are the principal form of reinforcement used because they offer a good combination of strength, stiffness and economy. Improved strengths and stiffness can be achieved with other fibres such as PA/aramid (Kevlar) or carbon fibres, but they are more expensive (in fact, there are also hybrid systems used to get a good balance of properties at an acceptable price). In composites, the matrix binds the reinforcing components and protects them from corrosion-oxidation or from other environmental effects. Glass fibres are usually sized and coated by different chemicals (such as, PVC, polyacetate, chrome chloride, polyvinyl acetate silane, or epoxy silanes), to help to improve their binding to the polymeric matrix.

Generally any remnants of monomers, catalysts and solvents that are left in the composite, can be the main cause of any health hazard to different extents and degrees. Otherwise, a properly prepared and finished ready to use composite product should normally not show any health hazard.

Thermoset composite compositions, prior to their final curing, are usually put in the shape of the final product, called *prepregs*. The prepreg approach is applied in manufacturing advanced composite systems, such as in the aircraft industry. For non-DGEBA epoxy resins, as well as TGPAP during the production of epoxy are found to cause sensitisation from their 'pre-pregs'.

Health hazards from plastic composites can be mainly due to contact (through skin) or due to inhalation.

As for the, hazards associated with skin contact, usually, dermatitis (allergic contact dermatitis) is the common occupational health complaint that has been reported for thermoset plastics materials. It is known that plastic composites based on polyester, epoxy and vinyl ester (epoxy diacrylate) resins can cause 'contact dermatitis'.

Common causes of irritation in general, are from the glass fibre reinforcement, (as well as from the dust from the product finishing processes). As far as glass fibres are concerned, for a long time it has been well documented that exposure to any airborne reinforcements (i.e., glass fibre) are mainly responsible for upper airway irritations, bronchitis, even lung cancer. The case of asbestos, although it is now banned and is not used as reinforcements, should be remembered. In one example, formaldehyde resin coated in glass wool was shown to cause serious immediate as well as delayed health hazards, even though no direct contact was involved (in this case, the truck who had been delivering glass wool for four years, after which he begin experiencing loss of voice and difficulty in breathing, where the upper mucosa was found damaged completely due to the glass fibres).

However, it is also known that, DGEBA epoxy resin, amine hardeners, dicarboxylic (phthalic) anhydride hardeners, and reactive epoxy diluents are all common causes of allergic contact dermatitis. And since all are health hazard causing organic pollutants, no remnants should be left in the composite, and must be decreased or avoided completely [73, 74].

Any unpolymerised epoxy left in the composite system are reported to be responsible for sensitisation and allergic contact dermatitis [69, 70].

4.2.3 Rubbers/Elastomers

Rubbers (elastomers, derived from 'elastic polymers') are 'highly' elastic materials that can be easily deformed by mechanical means. They, by definition, are materials capable of stretching and recovering their original shapes, after deformation to great extents. They are members of the polymer family consisting of flexible, long chain like, polymer molecules. In the, so called, green state, these flexible polymer molecules are not bound chemically to each other, and when deformed, the system cannot recover completely to its original state. To increase mechanical recovery and resilience, chain molecules are connected chemically by crosslinks (being similar in thermosets) by the curing (vulcanisation) process, after which, a fully and reversibly recoverable state is reached, and then the compound is called a rubber. Most rubber products are used as tyres, as well as for other mechanical parts (such as mountings, gaskets, belts, hoses).

Rubbers can be natural or synthetic such as, SBR, or polybutadiene (PBD).

Natural rubber latex (NRL) is obtained from rubber trees as a milky latex fluid (with small particles of rubber dispersed in an aqueous matrix), with about 1% of allergenic fraction of proteins (protein allergens like α-glubulins, hevein). From the concentrated suspensions of NRL, products like medical gloves, catheters, tubes and condoms are either produced directly (by simple dipping, extruding or coating techniques), or after its coagulation into dried or milled sheet forms.

NRL is an irritant because of the protein allergens and can cause allergic skin reactions associated with type I (immediate effect) hypersensitivities of 'hand dermotoses' and 'immunologic contact urticaria'. It is also possible to have type IV (delayed effect) hypersensitivities of NRL, due to the chemicals added during its processing (such as accelerators of the thiuram/carbamate/mercaptobenzothiazole types needed for vulcanisation, a number of different antioxidants/antiozonants, emulsifiers, extenders, colorants, retarders, stiffeners, biocides). Either of these types can be serious and even life threatening in some cases [75].

Synthetic latex (synthetic rubber) is produced industrially (i.e., SBR, SBS, SIS, PBD, polysilicones, ethylene-propylene-diene rubber and so on), and can be used alone or along with natural rubber in certain formulations.

The synthetic rubber industry uses a number of hydrocarbon additives, specifically called process oils (to act as a plasticiser, used below 20 phr) or extenders (used to keep the costs down). There are a wide range of mineral oils used as process oils, produced by blending of crude oil distillates and these may be either: paraffinic, naphthenic or aromatic. Process oils containing polycyclic aromatic hydrocarbons, are classified as potential carcinogens (and their use is decreasing considerably).

Most rubbers burn easily. Self-extinguishing properties can be obtained by appropriate compounding. Burning rubber produces considerably high heat and acrid smoke, which may contain harmful constituents, including halogens.

Nitrosamines and nitrosatable substances, which are potent carcinogenic chemical compounds, can exist in a number of latex and rubber products (i.e., balloons, gloves, toys, baby bottle teats, soothers and condoms), mainly because of certain accelerators used during their processing, and that can be released subsequently. A recent survey showed considerable amounts of release of these substances from most of the rubber toy balloons (with amounts above that of recommended level in Germany of 10 mg/kg material) as well as from rubber condoms (with amounts varying from < 10 to 660 mg/kg material, up to 1.4 mg nitrosamine per condom) [76]. The value mentioned in the latter, however, is not expected to be of toxicological significance [77]. EU legislation has limited the release of *N*-nitrosamines and nitrosatable compounds in teats and soothers to 0.01-0.1 mg/kg rubber.

Thermoplastic rubbers (TPR, or elastomers) are another group of materials that exhibit the same desirable characteristics as rubber, with the ease of processing of thermoplastics: at room temperatures they behave like a rubbery thermoset material. However, at higher temperatures, they behave like a thermoplastic (i.e., segmented block copolymers of PU, styrenics and polyesters, and olefinics.

4.3 Some Additional Notes on the Toxic Chemicals Evolving from Degradation, Combustion and Sterilisation of Polymers

Polymers with their long chain structures have a higher physical order than their monomers with much lower molecular weights, which lead them to decompose more easily under various environmental effects (like heat and UV light), and they are much more easily subjected to other deteriorating effects of chemicals and moisture. If stabilisers are not used, a rapid degradation of the material usually results, even under rather mild environmental conditions, throughout their shelf and service lives, meanwhile emitting some toxic volatile organic compounds in some cases and ultimately affecting their performance and lifetime.

On the other hand, when combusted (such as in fire), they can produce toxic gases that may seriously hamper the health.

4.3.1 On Toxics from Degradation of Polymers

Polymers, in the form of plastics and rubbers offer a number of benefits in use, but their production, application and disposal stages may present a number of environmental problems. One of these problems is 'degradation' due to either 'thermal, mechanical, UV or other' effects (i.e., in the case of a fire), and hence the production of toxic chemicals that can be taken by humans (by inhalation or by other means).

If lifetime stages of polymers are considered (which is the famous 'from cradle-to-grave issue' [78]), there can be a number of processes to consider, such as, melt degradation, long-term heat ageing and weathering, all based on different mechanisms (i.e., thermomechanical, catalytical, and radiation induced oxidations and scissions, as well as environmental biodegradation). The products evolving from the system can be due to the polymer as well as due to the additives, and they are usually low molecular weight, toxic gaseous molecules that emit from the system.

For health and safety purposes, the products that are likely to be formed should be known and identified precisely at all times.

If thermal (anaerobic) degradation is considered, there are two different mechanisms accepted: degradation occurs either via depolymerisation or via statistical fragmentation, and can follow three major pathways: side-group elimination, random scission or depolymerisation). For example, for PVC, in the first step of thermal degradation, elimination of side groups forms hydrogen chloride (gas) termed dehydrochlorination, that occurs up to 350 °C, and the remaining polyene macromolecules undergo further reactions forming a series of aromatic molecules such as benzene and toluene before further breaking up at 550 °C.

PVC degrades slowly on standing, with a colour change from light yellow to reddish brown caused by HCl evolution. Temperature affects the rate of this evolution but not its amount [78].

Thermal depolymerisation products are monomers (or co-monomers in the case of copolymer degradation), i.e., polymethacrylates and PS degrade by depolymerisation mostly - and monomeric-dimeric and trimeric units (methacrylates or styrenes, respectively), split off from the system. Both are suspected of causing cancer. For PS, monomer yield of thermal depolymerisation can be as high as 40%, while PMMA can decompose to the monomers completely above 300 °C [79, 80].

Polymers that do not depolymerise by heating, can in general decompose into smaller fragments by random scission, i.e., PE; which are bigger in size than monomeric units. This fragmentation can be initiated by thermal, as well as by chemical, mechanical or radiation means.

Thermal degradation of SBR rubber gives rise to monomers (1,3-butadiene and styrene) as well as to some toluene which turns into carbon dioxide, carbon , monoxide and water, at high temperatures.

The common gasket and O-ring material, nitrile-butadiene rubber, when degraded thermally, can give off butadiene and acrylonitrile gases, both of which are poisonous.

PU begins to degrade at rather low temperatures (200 °C) with a nitrogen containing yellow smoke and a solid residue. The yellow smoke then transforms into several nitrogen containing products, such as hydrogen cyanide and acetonitrile.

4.3.2 Toxic Compounds from Combustion, Thermo-Oxidative Degradation, Sterilisation and Others

US fire statistics show that over half of all fire deaths are from smoke inhalation, mostly of carbon monoxide and hydrogen cyanide. For the hydrogen cyanide, usually PU (mostly from foam in construction or furniture) are held responsible [81].

It is generally assumed that plastics are mainly responsible for production of hazardous emissions, if combustion of their waste is also considered. In addition to a number of toxic gases, heavy metals (cadmium, lead and so on) and the poisonous chemical products (such as dioxins and furans) produced, are all blamed on plastics. In a number of articles, it is shown that the formation of polychlorinated dioxins and furans are due to the combustion of PVC or its co-incineration with others.

Oxidative degradation occurs at much lower temperatures than conventional pyrolysis temperatures, and it is characterised by the complete random scission in the polymer chain backbone, compared to thermal degradation where polymer scission occurs randomly and at the chain end.

The thermo-oxidative degradation products of PMMA are those from monomer oxidation, as well as from oxidation of carbon monoxide and methane.

PP is known to be vulnerable to oxidative degradation at high temperatures and in sunlight.

For SBR, at lower temperatures oxidative degradation predominates, while thermal degradation takes over at higher temperatures.

Sterilisation of plastics packages in aseptic processing of foods as well as medical products is a rather common process, during which there is always the possibility of degradation and production of some new chemicals from plastics (such as a number of peroxides, alcohols, and a number of low molecular weight oxygen containing new compounds). Sterilisation is being mainly done with ionising radiation. For this, either accelerators (with up to 10 MeV) and ^{6}Co radionuclide sources (with 1.33 MeV gamma rays), as well as X-rays and ion beams are used [82]. In the US, Canada and EU, empty packaging plastics materials such as, laminated plastic films with aluminium foil are used for hermetically sealed products, dairy product packaging and single serving containers, food wrapping materials, co-extruded BOPP laminates and so on and are sterilised mainly by irradiation. The effects of irradiation on plastics is documented well, and it is known that interaction of ionising irradiation with plastics materials can lead to formation of certain new reactive intermediates, free radicals, ions and excited species; which manifests itself either in crosslinking (mostly for PE, PP and PS), or chain scission of the polymer through disproportionation, hydrogen abstraction, re-arrangement and formation of new bonds, as well as oxidative degradation [83]. During the sterilisation process, there is the possibility of production of certain low molecular weight (mostly volatile or migratable) radiolysis products from irradiated plastics packaging materials, which are of considerable interest. There are several packaging legislative requirements in some countries already, which necessitates the evaluation of compositional changes in the bulk plastic and evolution of volatiles after irradiation, in particular for polyolefins and PVC [84-86].

A recent standard specifies all aspects of sterilisation of medical plastics, which can be extended for all plastics as well, without any limit on the energy used for sterilisation [87].

On the other hand, there are also studies showing the effects of UV irradiation on the change of migration characteristics of the matrix polymer, (i.e., effect on the migration of phthalate ester plasticisers from clear PET water bottles [87]), and some data are provided on the effects of ionising irradiation on polymer additives, monomers and polymers themselves in general [88].

References

1. V.O. Sheftel and V.G. Sinitsky, *Gigiena I Sanitaria*, 1973, **3**, 111. (in Russian).

2. B.Y. Kalinin, L.P. Zimnitskaya and V.M. Zalesskaya in *Proceedings of the 2nd All-Union Conference*, Eds., A.A. Letavet and S.L. Danishevsky, Khimiya, Leningrad, Russia, 1964, p.110. (in Russian)

3. I.B Michailets, L.V. Sukhareva and V.I. Yevsyukov in *Environmental Protection in Plastics Production and Hygiene Aspects of Their Use*, Plastpolymer, Leningrad, Russia, 1978, p.99. (in Russian)

4. W.J. Uhde, and H. Woggon, *Nahrung,* 1976, **2**, 185.

5. R. Goydan, A.D. Schwope, R.C. Reid and G. Cramer, *Food Additives and Contaminants*, 1990, 7, 3, 323.

6. V.O. Sheftel, *Hygiene Aspects of the Use of Polymeric Materials in the Water Supply,* All-Union Research Institute of Hygiene and Toxicology of Pesticides, Polymers and Plastic Materials, Kiev, Russia, 1977, p.61 (in Russian). [PhD Thesis]

7. C. Vasile and M. Pascu, *Practical Guide to Polyethylene*, Rapra Technology Ltd., Shawbury, Shrewsbury, UK, 2005.

8. *Healthy Building Network*, www.healthybuilding.net

9. G.K. Lemasters, A. Hagen and S.J. Samuels, *Journal of Occupational Medicine*, 1985, **27**, 2, 490.

10. N.S. Zlobina, A.S. Izjumova and N.J. Ragul'e, *Gigiena Truda i Professional'nye Sabolenavija*, 1975, **12**, 21.

11. J.M. Norris in *Proceedings of the SPI Reinforced Plastics/Composites Institution, 40th Annual Conference*, Atlanta, GA, USA, 1985, Paper No.22-C.

12. H. Voss, *Kunststoffe*, 1987, **77**, 1, 23.

13. J.V. Rutkowski and B.C. Levin, *Report for the National Bureau of Standards (NEL)*, Gaithersburg, MD, USA, 1985, **85**, 3248, p.60.

14. B.J. Dowty, J.L. Laseter and J. Storer, *Pediatric Research*, 1976, **10**, 7, 696.

15. *Chemical Industy Archives*, http://www.chemicalindustryarchives.org/dirtysecrets/vinyl/1.asp

16. Norsk Hydro, *PVC and the Environment*, Norsk Hydro, Oslo, Norway, 1996.

17. *Dioxin: Seveso Disaster Testament to Effects of Dioxin*, http://www.getipm.com/articles/seveso-italy.htm

18. *Compilation of EU Dioxins Exposure and Health Data. Task 2: Environmental levels*, http://ec.europa.eu/environment/dioxin/pdf/task2annex.pdf

19. K. Mulder and M. Knot, *Technology in Society*, 2001, **23**, 2, 265.

20. N.M. Fayad, S.Y. Sheikheldin, M.H. Al-Malack, A.H. El-Mubarakv and N. Khaja, *Journal of Environmental Science and Health*, 1997, **32**, 4, 1065.

21. *US Federal Register*, 1998, **63**, 83, 23785.

22. *The Dioxin Homepage, Environmental Justice Activists*, www.ejnet.org/dioxin/

23. *Toxicological Profile For Di(2-Ethylhexyl) Phthalate*, US Deparrtment of Health and Human Sciences, 2002, www.atsdr.cdc.gov/toxprofiles/tp9.pdf

24. R. Hokanson, W. Hanneman, M. Hennessey, K.C. Donnelly, T. McDonald, R. Chowdhary and D.L. Busbee, *Human and Experimental Toxicology*, 2006, **25**, 12, 687.

25. L. Olie, L-G. Hersoug and J.O. Madsen, *Environmental Health Perspectives*, 1997, **105**, 9, 972.

26. C-G. Bornehag, J. Sundell, C.J. Weschler, T. Sigsgaard, B. Lundgren, M. Hasselgren and L. Hägerhed-Engman, *Environmental Health Perspectives*, 2004, **112**, 14, 1393.

27. C.J. Howick in *Proceedings of PVC'99*, Brighton, UK, 1999, p.233.

28. *Plastics News*, USA, 2000, **12**, 8, 19.

29. ENDS Report, 1997, **266**, 11.

30. H. Arvela, *Science of the Total Environment*, 2001, **272**, 1-3, 169.

31. *ENDS Report*, 1997, 266, 35.

32. *Vinyl 2010*, www.vinyl2010.org

33. *Methyl Methacrylate (MMA)*, Hazard Evaluation and Information Service, http://www.dhs.ca.gov/ohb/HESIS/mma.htm

34. T.G. Sharova, *Gigiena Truda i Professional'nye Zabolevaniya*, 1989, **3**, 12.

35. *Regulations (Standards – 29CFR) Acrylonitrile – 1910.1045*, US Department of Labour, www.osha.gov/pls/oshaweb/owadisp.show_document?p_id=10065&p_table=standards

36. K. Hashimoto, *Sangyo Igaku*, 1980, **22**, 5, 327-47.

37. T. Colburn, D. Dumanoski and J. Myers, *Our Stolen Future,* Myers Publications, Dutton, NY, USA, 1997.

38. *Chemical in plastics can lead to several health problems*, News Target, http://www.newstarget.com/007131.html

39. *European Commission, Opinion of the Scientific Committee on Food on Bisphenol A*, http://europa.eu.int/comm/food/fs/sc/scf/out128_en.pdf

40. F.S. vom Saal and C. Hughes, *Environmental Health Perspectives,* 2005, **113**, 8, 926.

41. K. Sissell, *Chemical Week*, 2004, **166**, 29, 28.

42. S. Eberhartinger, I. Steiner, J. Washüttl and G. Kroyer, *Zeitschrift für Lebensmittel Untersuchung und Forschung*, 1990, **191**, 286.

43. S. Monarca, R. De Fusco, R. Biscardi, V. De Feo, R. Pasquini, C. Fatigoni, M. Moretti and A. Zanardini, *Food and Chemical Toxicology*, 1994, **32**, 9, 783.

44. H. Kim, S.G. Gilbert and J.B. Johnson, *Pharmaceutical Research*, 1990, 7, 2, 176.

45. W. Shotyk, M. Krachler and B. Chen, *Journal of Environmental Monitoring*, 2006, **8**, 2, 288.

46. *WHO risk assessment for Sb in the drinking water*: http://www.who.int/water_sanitation_health/dwq/chemicals/antimonysum.pdf

47. T.H. Begley, J.L. Dennison and H.C. Hollifield, *Food Additives and Contamination*, 1990, 7, 797.

48. G. Linssen, H. Reitsma and G. Cozynsen, *Zeitschrift für Lebensmittel Untersuchung und Forschung*, 1995, **201**, 253.

49. S. Eberhartlinger, J. Steiner, J. Washuttl and G. Kroyer, *Zeitschrift für Lebensmittel Untersuchung und Forschung*, 1990, **191**, 286.

50. H. Piekacz, *Roczniki Panstwowego Zakladu Higieny*, 1971, **22**, 295. (in Polish)

51. M.T. de A Freire, L. Castle, F.G. Reyes and A.P. Damant, *Food Additives and Contaminants*, 1998, **15**, 4, 473.

52. M. Kashtock and C.V. Breder, *Journal of Association of Official Analytical Chemists*, 1980, **63**, 168.

53. P.Y. Pennarun, P. Dole, and A. Feigenbaum, *Journal of Applied Polymer Science*, 2004, **92**, 5, 2845.

54. G. Sadler, D. Pierce, A. Zervos and S. Katschke in *Proceedings of the SPE Philadelphia Section 1996 Regional Technical Conference (RETEC) Regulatory Issues in the Plastics Industry: Health, Safety and the Environment*, Cherry Hill, NJ, USA, 1996, p.45.

55. V.O. Sheftel, *Indirect Food Additives and Polymers: Migration and Toxicology*, Lewis Publishers, Boca Raton, FL, USA, 2000.

56. *Chemical Week*, 2005, **167**, 13, 27.

57. S. Alex, *Chemical Week*, 2005, **167**, 17, 31.

58. D.R. Tierney, T.R. Blackwood and, G.E. Wilkins, *Status Assessment of Toxic Chemicals: Acrylonitrile*, Environmental Protection Agency, OH, USA, 1979, p.43.

59. H.C. Park and R.C. Ashcraft, *Journal of Plastic Film and Sheeting*, 1985 **1**, 2, 95.

60. *Perfluorooctanoic Acid Human Health Risk Assessment Review Panel*, EPA Science Advisory Board, http://www.epa.gov/sab/panels/pfoa_rev_panel.htm

61. J. Eilperin, *Washington Post*, 2006, January 26th, p.A01.

62. *DuPont Subpoenaed In Teflon Ingredients Health Probe*, Bloomberg.com, www.dupontcouncil.org/dupont_subpoenaed_in_teflon.htm

63. W. Hemmer, M. Focke, F. Wantke, M. Götz and R. Jarisch, *Journal of the American Academy of Dermatology*, 1996, 35, 3, Pt 1, 377.

64. R. Rajaniemi, *Occupational Medicine* 1986, **36**, 2, 56.

65. *NPI Substances- Health and Environmental Standards and Guidelines*, http://www.npi.gov.au/database/substance-info/sources.html

66. R.A. Plastinina, G.V. Pavlova, N.A. Oleinik, V.I. Oshchepkov and I.A. Shinkareva *Gigiena Truda i Professional'nye Zabolevaniya*, 1986, **12**, 16. [in Russian]

67. J.R. Nethercott, H.R. Jakubovic, C. Pilger and J.W. Smith, *British Journal of Industrial Medicine,* 1983, **40**, 3, 241.

68. G.E. Hartzell, *Journal of Cellular Plastics*, 1992, **28**, 4, 330.

69. G.E. Hartzell in *Proceedings of the SPI Polyurethanes World Congress: The Voice of Advancement*, 1991, Nice, France, p.40.

70. E.M. Abdel-Bary in *Polymers In Construction,* Ed., G. Akovali, Rapra Technology Ltd., Shawbury, Shrewsbury, UK, 2005.

71. A.P. Yavorovsky and V.S. Bogorad, *Gigiena Truda i Professional'nye Zabolevaniya*, 1989, **3**, 28. (in Russian)

72. K. Tarvainen and L. Kanerva, *Journal of Environmental Medicine*, 1999, **1**, 1, 3.

73. L. Kanerva, R. Jolanki, J. Toikkanen, K. Tarvainen and T. Eastlander in *Irritant Dermatitis*, Eds., P. Elsner and H.I. Maibach, Current Problems in Dermatology Series, Volume 23, Karger, Basel, Switzerland, 1995.

74. T. Fischer, S. Freoert, I. Thulin and L. Rulsson, *Contact Dermatitis*, 1987, **16**, 1, 45.

75. C. Vincenzi, N. Cameli, A. Vassilopoulou and A. Tosti, *Contact Dermatitis*, 1991, **24**, 2, 66.

76. G.K. Lemasters, A. Hagen and S.J. Samuels, *Journal of Occupational Medicine*, 1985, **27**, 7, 490.

77. C.P. Hamann, *American Journal of Contact Dermatitis*, 1993, **4**, 4.

78. W. Altkofer, S. Braune, K. Ellendt, M. Kettl-Grömminger and G. Steiner, *Molecular Nutrition and Food Research*, 2005, **49**, 3, 235.

79. E. Proksch, *International Journal of Hygiene and Environmental Health*, 2001, **204**, 2, 103.

80. K. Pielichowski and J. Njuguna, *Thermal Degradation of Polymeric Materials*, Rapra Technology, Shawbury, Shrewsbury, UK, 2005.

81. Y. Ito, H. Ogasawara, Y. Ishida, H. Ohtani and S. Tsuge, *Polymer Journal (Japan)*, 1996, **28**, 12, 1090.

82. A.S.Wilson, *Plasticisers Selection, Applications and Implications*, Rapra Review Report, 1996, **8**, 4, Report No.88.

83. S. Toloken, *Plastics News*, USA, 1999, **11**, 19, 1.

84. N.H. Stoffers, J.H. Linssen, R. Franz and F. Welle, *Radiation Physics and Chemistry*, 2004, **71**, 1-2, 205.

85. R.L. Clough, *Nuclear Instruments and Methods in Physics Research Section B*, 2001, **185**, 1-4, 8.

86. P.G. Demertsiz, R. Franz and F. Welle, *Packaging Technology and Science*, 1999, **12**, 3, 119.

87. J. Messinck in Proceedings of SPE Polyolefins XII, Conference, Houston, TX, USA, 2000.

88. EN ISO11137, *Sterilisation of Health Care Products - Radiation*, 2006.

Some Additional References

1.   W.J. Uhde and H. Woggon, *Nahrung*, 1976, **2**, 185.

2.   L. Castle, A. Mayo and J. Gilbert, *Food Additives and Contaminants*, 1989, **6**, 4, 437.

3.   M.G. Melian, E.B. Torres and V. Diaz, *Revista Cubana de Higiene y Epidemiologia*, 1985, **23**, 4, 441.

4.   *The Styrene Information and Research Center (SIRC)*, www.styrene.org

5.   *Environmental Justice Activists, The Polystyrene Pages*, http://www.ejnet.org/plastics/polystyrene/

6.   *Chemical Industry Archives: Vinyl Chloride*, http://www.chemicalindustryarchives.org/dirtysecrets/vinyl/1.asp

7.   F.E. Borrelli, P.L. de la Cruz and R.A. Paradis, *Journal of Vinyl and Additive Technology*, 2005, **11**, 2, 65.

8.   M.S. Brown in *Controversy: Politics of Technical Decisions*, Ed., D. Nelkin, 3rd Edition, Sage Publications, Newbury Park, CA, USA, 1992.

9.   *DuPont's Official web site, 'Teflon'*, www.dupont.com

10. *Polymer Science Learning Centre*, PTFE, http://pslc.ws/mactest/ptfe.htm

11. *Some Monomers, Plastics and Synthetic Elastomers and Acrolein*, IARC Monographs on the Evaluation of Carcinogenic Risks to Humans, Volume 19, IARC, Lyon, France, 1979, p.285.

12. R. de Fusco, S. Monarca, D. Biscardi, R. Pasquini and C. Fatigoni, *The Science of the Total Environment*, 1990, **90**, 241.

13. M. Kashtock and C.V. Breder, *Journal of Association of Official Analytical Chemists*, 1980, **63**, 2, 168.

14. P.J. Fordham, J.W. Gramshaw, H.M. Crews and L. Castle, *Food Additives and Contamination*, 1997, **12**, 5, 651.

15. A.I. Bazanova in *Proceedings of the 3rd All-Union Conference*, Ed., L.S. Danishevsky, Khimiya, Moscow, Russia, 1966, p.113. (in Russian).

16. P. Blagoeva, I. Stoichev, R. Balanski, L. Purvanova, T.S. Mircheva and A. Smilov, *Khirurgiia*, 1990, **43**, 6, 98. (in Bulgarian).

17. *Food Safety - from the Farm to the Fork, EU Commission*, http://europa.eu.int/comm/food/food/chemicalsafety/foodcontact/index_en.htm

18. *The Polyurethanes Book*, Eds., D. Randall and S. Lee, Wiley, London, UK, 2002.

19. P. Penczek, P. Czub and J. Pielichowski in *Crosslinking in Materials Science*, Advances in Polymer Science Series, Volume 184, Springer-Verlag GmbH, Berlin, Germany, 2005.

5 Plastics as Food and Packaging Materials, Rubbers in Contact with Food, and their Possible Health Effects

5.1 Introduction

Certain chemical components of plastic food packages or rubber parts that are contacting with food can migrate and transfer into the latter, and can became a component of the food [1].

Since the polymer itself, which constitutes the main part of the plastic or rubbery materials, is composed of big molecules (with high molecular weights), they are not expected to migrate into food. However, a number of additives that exist in the system, and residual monomers and oligomers that are much smaller with low molecular weights, can migrate more easily into the food. Migratory molecules, which can be additives, remnants of monomer, or other residues possibly left from the polymerisation as outlined in the previous chapter (these may be catalyst residues, remnants of chain agents and so on). Depending on the type and amount of these migratory molecules (or contaminants), toxicological data must be provided.

Whether or not a migrating molecule produces unwanted effects certainly depends on its composition in the material originally, and in the food after migration. There are studies showing that the type of the polymer material itself, time of contact and type of migrating element affects the migration process at large [2].

Migrating molecules must be safe at the levels that they are expected to be found in the food. These levels need to be determined by sensitive analytical methods, since these amounts are extremely small, at the level of parts per million/ppm, or parts per billion/ppb.

The US FDA has permitted substances that contain low levels of toxic impurities to remain in the system, when the amounts expected to become a component of food have been found to be of no toxicological significance (*de minimis*), called the *de minimis* policy [1].

Some researchers have suggested that even minute amounts of certain chemical compounds can act directly or may adversely change the way humans and wildlife develop and reproduce. Their effects can be important and even vital, and they may be the reason for a number of health problems encountered, anything from the long-lasting allergy or asthma or even a lasting headache, to more serious issues, such as cancer.

It should be noted, however, that use of plastics or rubbers in various applications where contact with food is involved should not automatically mean that its use will result in adverse health effects. A number of plastics and rubbers are being used in many critical areas already without any problems, such as food packaging materials and in the health sector (i.e., blood bags and dialysis tubing). In fact, in most cases, it is not the plastic or rubber material itself, but certain additives and other foreign chemicals (that are added for different purposes or that are produced in the systems somehow), that can be the source of the problems that can pose health hazards, hence they should be considered carefully.

This being the case, the same plastic or rubber material can be very safe or very hazardous to humans - depending on the ingredients in it.

Plastics are mainly used as 'packaging material', as mentioned before, which can range from shopping bags to various containers, and they are mostly used in food packaging. A food packaging material means a direct contact with the food, for shorter or longer times and at temperatures changing from sub-ambient to well above ambient, including the applications of microwave ready foods contained in plastic film packages.

Rubbers are rarely used in food packaging, with the exception of rubber seals (in beer bottles and jars), and various leap seals (in food cans). In the food industry, on the other hand, there are a number of application areas where rubbers can come in contact with food (like pumps, gaskets in exchangers and machinery, general seals in machinery and storage vessels, and so on).

Food contact approval, in general, is in compliance with European Directive 2002/72/EC, specifically for use in all finished, film-type food packaging materials.

5.2 Outline of Plastics Packaging and Possible Health Effects Involved

For the packaging materials, in general: the contents (protection and safety of the contents inside the package), and the environment, are both important for protection, and any packaging material is supposed to provide optimum safety both to the goods inside the package and to the environment outside.

A proper plastic food packaging material must function properly as the protection of the food is considered is important to:

(i)   prevent the contents from any possible spoilage and contamination, polluting agents or leakages, and,

(ii) keep the freshness of contents as long as possible - by providing appropriate controlled atmospheres (controlled atmosphere packaging (CAP)) inside, and even oxygen barriers outside of the package, whenever needed.

For the safety of the environment, it is mainly of interest to prevent leakage of possible pollutants, toxics and aggressive products from the package by providing, for example, a shatter resistant package, or by providing additional safety such as safety caps, child resistant caps.

The packaging market is the biggest sector of plastics consumption, where roughly one-third (around 50 million tonnes per year) of the total plastics produced is used worldwide as mentioned previously, and this consumption is still growing continuously. About one-quarter of the plastics are used to produce the plastics packaging film and sheets, which are everywhere in our daily life: wrapping of food, office supplies, clothing, bags and sacks. Within these, since they come directly in contact with the food, plastics food packaging materials are of utmost importance as regards health concerns. All flexible or semi-rigid food packaging plastics need to be treated with the utmost care, especially microwave-safe, food grade plastics packaging materials, which are being transferred directly from freezer to the oven, then to the table. Since plastics packaging material can be either in a simple mono or a complex multi-layered structure (and may be prepared with adhesive-bonded layers in the latter case), with brands and descriptions printed on top of it, each of these can affect health and should be considered separately, hence, a number of complex factors have to be accounted for.

To give a better picture of the state-of-the-art, a rather more descriptive presentation will be made in the next section.

5.2.1 Why Plastics in Packaging?

Plastics are used preferentially in packaging, because they have high performance to weight ratios, they offer considerable ease of processing, and they have some gas barrier properties providing easy moisture proofing and sealability. In addition, they are light in weight and can offer an exceptional combination of properties when combined with other materials (i.e., paper and glass). Their chemical resistances (to acids, bases, organic chemicals) and inertness to facing contents (particularly to food products) is an advantage. They are easily sterilised, and they are aesthetically pleasing. Plastics used as packaging materials have already opened up new horizons to help provide a much higher comfort level in our lives.

Plastics packaging is already a multi-billion dollar huge business in the US alone.

5.2.2 Types of Plastics Used in Packaging

Thermoplastics, in particular the commodity plastics, polyethylene (PE), polypropylene (PP), polyvinylchloride (PVC) and polystyrene (PS), are the most commonly used in packaging (over 90% of packaging is thermoplastics, followed by a small amount of thermosets, composites, rubber and thermoplastic elastomers (TPE).

a) **Thermoplastics:** Within the thermoplastics used, polyolefin use is the highest (PE and PP; about 65%), followed by polyethylene terephthalate (PET), PVC and PS. Other thermoplastics, such as polyacrylates (i.e., polymethylmethacrylate (PMMA), polyamides (PA) and polycarbonates (PC) are used for specific applications in plastics packaging.

b) **Thermosets:** Polyurethanes (PU) and other thermosets (about 1%) and composites are used as packaging materials, mainly glass fibre reinforced polyesters and reactive reaction injection moulded items (less than 1%) are used for certain limited exceptional applications.

c) **Others:** There are also other types of packaging films and packaging seals made out of a number of different versions of TPE, such as copolyester (COPE) (which are TPE-E triblock copolymers of high performance TPE), polyether block amide, olefin-based TPE and PP/ethylene-propylene diene terpolymer-thermoplastic vulcanisates.

5.2.3 Types and Forms of Plastics Packaging

The main aim of food packaging materials is to protect the contents from environmental effects, as well as to protect them from moisture, drying-out and the effects of oxygen. Nowadays it is possible to have more airtight plastic packaging systems using new nanoparticle technology (thus in the fridge when a cheese package is put next to salami, their smells will not be transferred).

There are different types of plastics packaging used in different applications.

These can be either (a) **common film structures** (which can be either in the common monolayer or in series of multilayered-coextruded film types) or, (b) **hybrids** (hybrid structures can be of plastic-to-metal or plastic-to-paper type, i.e., multilayer films with paper and with metal foils, protective coating for metals, and so on).

Plastic films are produced for their use either in **packaging** (which are either for food or non food, or for industrial/agricultural) or **non-packaging** applications.

Within these, food packaging applications will be our main interest.

There may be different forms of plastics packaging, as follows:

a) **Packaging films** (mainly made of PE, polyester, PP and PVC),

b) **Containers** (bottles, cans and others, mostly made of PVC, PET, and PE),

c) **Polymeric foams** (used mainly for damping and protective packaging, mostly made of PS, PE, PU and others), and

d) **Structural packaging** (rigid and transparent plastic sheets made mostly of PVC, PC and PMMA that are used for blister packs and boxes).

Examples of several different groups of plastics used as packaging film materials are:

(i) **Polyolefins** (the most widely used group in packaging, mainly PE and its copolymers, PP, and its ionomers),

(ii) **Styrenics** (mainly PS, styrene-butadiene rubber (SBR), expanded (foamed) polystyrene (EPS)),

(iii) **Vinyls, vinylics** (mainly PVC, polyvinylidene chloride (PVDC),

(iv) **Polyesters** (mainly PET, PC),

(v) **Engineering plastics** (mainly PC, PET, PA, polyether imide, polysulfone, liquid crystal polymers of PET),

(vi) **Barrier polymers** (mainly PVDC polymer and copolymers, ethylene vinyl acetate (EVA), ethylene–vinyl alcohol copolymers (EVOH) and PA).

Plastic films are made mostly of commodity plastics (i.e., PE, PS, PVC, PP; in the EU about 70% of plastic packaging films used are different PE) or engineering plastics.

5.2.3.1 Plastic Films and (Flexible) Plastics Food Packaging Materials and Related Possible Health Effects

There are four different types of plastic film applications:

a) In food packaging (in-store and confection bags, bags-in-a-box (wine), edible bags - for food directly to reheat in boiling water),

b) In non-food packaging (liners for shipments, sacks, bubble packing/envelopes,

c) In stretch/shrink wraps (strong, flexible films of linear low-density polyethylene (LLDPE)), low-density polyethylene (LDPE)), and PVC for domestic (stretch cling films) or of LLDPE, LDPE or PP films mostly industrial (heat-shrink wrap) uses,

d) In non-packaging (agricultural/construction barrier films, protective cloths, medical and health care films, bags).

Plastic films can be of:

(i)   'conventional/stretch/shrink' types,

(ii) 'monolayer/multilayer/oriented',

(iii) 'extruded/coextruded/cast' if prepared by extrusion or casting, and

(iv) with 'monomaterial/multimaterial/hybrid' structures.

In most of the cases, the permeation of oxygen and water vapour (for packaging applications) and CO_2 (for carbonated drinks applications) are very important. Since no single plastic offers the proper combination of price, degree of permeability and printability required for the application, multiple layers of (multilayered-coextruded) plastic systems are used, hence the required permeation constants for the composite layered wall may be maintained [3]. When several polymers are combined in the form of multiple thin layers (as is the case for multilayer/coextruded/hybrid systems), several functionalities can be integrated in such a multilayer-coextruded-multimaterial packaging system (i.e., layers of PE as the water barrier and polyvinyl alcohol (PVOH) as oil barrier), can all be combined as well. PA-6 layers are usually used to provide enhanced oxygen permeabilities when needed, while less oxygen permeable EVOH layers are preferred as a packaging material for sensitive products (i.e., to avoid the spoiling effect of oxygen on the fat of meat and cheese), and a different number of barrier plastic layers can be also combined as needed. However, for multi-layered structures, one should keep in mind that, there may be another complex problem of 'multiple compound migration', (i.e., it is possible for migration to occur from different layers, because each layer can have different additives). So the chemical characteristics of each of the different layers as well as the adherent layers must all be considered and checked carefully for the safety of the complete layered film structure of such systems. Because each plastic film layer may contain various additives, such as phthalates, bisphenol-A, and nonylphenols, that are usually present as plasticisers (used to make them (more) flexible and durable), or others. These contaminating chemicals can leach out into the foods they are in contact with, and as well they can evaporate into the gas phase, and can be inhaled. Increase in temperature usually speeds up both of these (microwaving foods in plastic packaging is usually discouraged).

In **Table 5.1** are listed some possible toxins, endocrine disruptors (ECD) and carcinogens that can migrate from plastic packaging materials to the food contained in the package.

In this context, another polymeric material that has (and is) been used in packaging should also be cited. This material is the 'paper', which is made of natural polymer (not a plastic) cellulose, which is an inert chemical. Although paper is inert, it is still claimed that the coating on the paper used to give grease resistance to microwave popcorn bags, fast food and candy wrappers, as well as in pizza box liners, can leach out certain toxic chemicals to the food in contact with it. When ingested (with the food), these chemicals are shown to break down into perfluoro-octanic acid, a chemical that an EPA expert

Table 5.1 Some possible toxins, ECD and carcinogens that can migrate from common plastic packaging materials to some food contents		
LDPE	Plastic bags, various food storage containers	Possible contaminants migrating: various antioxidants
HDPE	Plastic bags, yoghurt cups, milk jugs	Possible contaminants migrating: various antioxidants
PVC	Water/cooking oil bottles, meat wraps	Possible contaminants migrating: plasticisers (mainly phthalates), various stabilisers and heavy metal ions
PET	Soda/water bottles, beer cans, food jars	Possible contaminant migrating: acetaldehyde
PP	Bottle screw-caps, drinking straws	Possible contaminants migrating: butylated hydroxyl toluene (BHT), various stabilisers
PS	Meat trays, take-out food containers and cups	Possible contaminant migrating: styrene (accumulates in body fat)

panel last year found to be likely a human carcinogen [4]. In addition, di-butyl phthalate plasticiser can be present in some food-contact papers.

Plastic films have some handicaps as regards their environmental considerations and the negative public image they have, mostly due to their visually compelling appearance when accumulated in the environment as waste. Plastic bags are already restricted in use or taxed in countries such as Ireland, Germany, South Africa, Taiwan and Hungary, mainly due to the concerns about their disposal and environmental problems. Hence, studies to develop properly degradable polymers are a very active area of research for packaging materials. However, so far, studies for the use of blends of thermoplastics with starch and its derivates; and use of polylactic acid, polyglycolic acid, polycaprolactone, polyhydroxyalkanoate and polyhydroxybutyrate by themselves as degradable packaging materials drew considerable attention, but the outcomes are not finalised yet, mainly due to either their non-complete degradation in nature or concern on the health issues for the degradation fragments themselves.

Plastic packaging films are usually used with coloured and heavily printed surfaces, applied for creation of visual attraction to the contents. The characteristics of these coloured structures and prints on the film surfaces should also be considered carefully, because there is a high possibility that their contents, which can be toxic, may migrate through the thin film structure inwards to contaminate the food.

Thus, not only the contents of the plastic packaging films, but characteristics of the printing on it, should be considered and checked carefully [5].

More information on this topic is provided in Chapter 8.

5.2.3.3 Some Recent Developments for Additives Used in Food Packaging Plastics

A wide range of additives are used to enhance the performance, appearance and processabilities of food packaging materials, and legislation for the products in contact with food are continuously under review, throughout the world, which is applied to all of the materials in the compound (stabilisers, pigments, processing aids, lubricants, and so on). In this context, toxicity of plastics additives is well researched and documented. However, there has been an argument remaining as to which is better to legislate for, either:

a)   The inherent content of a plastic compound, or,

b)   The extractability of potentially hazardous substances.

The first statement is the one that is to be followed, however, if this approach is accepted, glass should be excluded as a packaging material since it contains lead, though this lead cannot be extracted. Anyway, a new legislation is on its way. For many years the effective international control of additives has been done by the US FDA, and most plastics and additives are tested according to these standards.

In the EU, there is extensive national legislation (the most influential body mention is the German Federal Health Ministry, Bundesgesundheitsamt - BGA, which is frequently used in material specifications).

The legislation of the EU goes back to the **1990 Directive-90/128/EEC**, which was originally a positive list of authorised monomers. The Commission published the list of additives that will require testing for migration in food-contact applications, in a new Directive, and the process of listing restricted additives will be continued by means of amendments to the 1990 Directive. The programme uses the amendments with some new concepts (such as food consumption factors, migration modelling and functional barriers). Otherwise, it is almost impossible to present an exhaustive list of additives that have been accepted by the various authorities for their use in contact with food.

The **EU Directive 2002/72/EC** regulates plastics intended to come in contact with food, where the principle of regulation is the establishment of a positive list, which presently is restricted to monomers and start substances. **Directive 2002/72/EC** (and its amendment **2004/19/EC**) contain in their Annexes I-IV, lists of evaluated monomers and additives and the corresponding specific migration limit (SML) values. SML is expressed as the amount of a substance (mg) allowed to migrate from the plastic into 1 kg of food. The law requires compliance with the SML for the end-use material, in addition to the requirement of compliance with the overall migration limit (60 mg/kg food).

Currently, for additives, there is an 'incomplete list' existing because the harmonised evaluation of all additives is not yet finalised. It is to be noted that for manufacturing food contact plastics, only positive list substances are allowed as additives [6].

General guidelines and some recent developments are summarised next:

Stabilisers/Antioxidants: Phosphite/phosphonites are generally regarded as the most effective stabilisers during processing, protecting both the polymer and the primary antioxidant. The most expensive stabilisers are organotin stabilisers, while lead compounds are the cheapest.

For applications in contact with food, FDA and BGA regulations recommend liquid antioxidants based on Vitamin E, developed as patented systems.

Among recent developments, there is the stabiliser of high-purity tris-nonylphenyl phosphite, with 0.1% residual nonyl phenol, which is FDA-approved for food-contact (and medical) applications.

There is also a new solid phosphite antioxidant developed by GE Specialty Chemicals, based on butyl ethyl propane diol chemistry, which has FDA approval for food contact applications.

Plasticisers: Polymeric plasticisers - usually polyesters, based on adipic acid are preferred for food contact applications - which have very low level of migration.

Stearic acid esters are used as plasticisers, processing agents and also as lubricants (for PS), with general food-contact approval.

Sebacates and adipates - provide good low temperature plasticisation, with fairly general food-contact approval. Di-butyl sebacate is a highly efficient primary plasticiser for low temperature applications, used in films and containers for packaging.

Epoxidised grades (soya bean and linseed oil) are used as stabilised plasticisers with no migration. Soya bean versions have widespread approval for food contact applications.

Colorants: Colorants can be critical as far as their possible extractions and toxicities are concerned. As for pigments, azoic yellows are known to be safe and suitable for food contact applications. New mixed-phase rutile yellow pigments also satisfy a number of food-contact requirements. Novel blue-red azo can be used as an alternative to high performance organics, and meets FDA requirements.

Thermochromic and *photochromic pigments* are micro-encapsulated liquid crystal systems, giving precise colour changes at specific temperatures (or when exposed to light), and are particularly interesting for food/pharmaceutical packaging, as they can give an indication of storage or cooking state. Most of the pigments in use comply with FDA food contact regulations.

'Intelligent' heat protection for food products via pigment systems can be incorporated in food packaging to control the temperature of heat-sensitive products, which is under study by WHO.

Pigment dispersants are low molecular weight ionomers that can promote good pigment dispersion and come within the regulations of many countries for colour concentrates in food contact applications.

Processing aids, lubricants and antistats: Most **fluoropolymer processing aids** (reduce die build-up and improve output of film) comply with indirect food contact regulations and can be used in PP and PE.

Nucleating Agents: Milliken's nucleating technology for PP is approved for food contact by the FDA, the Canadian Health Protection Board and the BGA.

GE fine particle silicone anti-blocking additive (used to improve film clarity and abrasion resistance) is approved for food contact.

As for the information sources on the topic, there is information available from national plastics federations and associations (i.e., the UK packaging and printing research association, Pira), as well as from the EU. A comprehensive database on all forms of chemical legislation in 25 countries has been published as a CD-ROM by the United Nations Economic Commission for Europe (UN/ECE) [7], which covers 15 sectors of the industry, including materials in contact with foodstuffs, with over 600 text summaries given along with full references. The database can be searched by country, keyword, and reference to act or date, with summaries and titles in the original language (as well as in English). In the Directive, there is also a database, which provides useful information for countries worldwide (including those that still have little or no legislation controlling chemicals), and some instant access information.

5.2.3.2 Plastic Bottles, Containers and Other Plastic Moulded Items

In this sector of plastics packaging, the consumption of various PE materials is the highest, followed by PET and finally there is the PVC (more in EU, and less in North America). Within these, the application of PET is increasing, because PET has several advantages over PE and PVC: PET is the most suitable for hot-filling, it is microwave proof and it is clear. PET has a better balance of gas permeabilities (better by combining the moisture impermeability of PE) and PET has a higher oxygen and carbon dioxide impermeability than PVC. In addition, there are new generation PET grades and blends available (i.e., new PET/polyethylene naphthanate blends or copolymers are available with aesthetic and performance characteristics similar to glass and the durability and convenience of cans, and an epoxy-amine coat barrier is being applied to further improve the impermeabilities).

5.2.3.3 Plastic Foams and Rigid Plastic Packaging

The basic principle for functions of packaging is applicable for this group of materials. Plastic foams and rigid packaging materials are used either:

a) To preserve the packed contents from environmental effects (by protecting them from mechanical damage, despoiling, or environmental contamination), or,

b) To protect the environment from the contents (from leakage of poisons, toxins and aggressive products packed inside them), or by offering a structural barrier (rigid packing) and protection against vibration and impacts (foam).

The term 'rigid packaging' covers a broad range of items from blisters (blister packs) to tubes and megabins (715-755 litres). The end uses of foam and rigid packing can change from 'high resistant or isothermal packaging, hybrid packaging, display and hygiene-medical packaging' to 'electronic and transport packaging'.

There is a wide range of different types of plastics that are being used in this sector:

a) Commodity plastics: within these, PE and PS are the most used in bulk and (to a lesser degree) in foamed forms, followed by PP (both in bulk and in foamed forms). PVC is preferred for blister packs.

b) Foams: in addition to PE, PS and PP, PU foams are also used.

c) Engineering plastics: within engineering plastics, there are acrylonitrile-butadiene-styrene, PET, PA, and PC (for its high clarity and high impact strength), reactive injection moulding PU, as well as a number of composites.

5.2.4 Smart Packaging

To complete the plastics packaging part, although it is not directly related to health issues, we want to include the recent concept of smart packaging, which is packaging that does more than protecting the contents. Smart packaging can be electronic, electrical, mechanical or chemical, i.e., a plastic pot of drugs that reminds the user when to take them and records what was taken and when. Smart packaging can indicate if certain pathogens are present in the packaging, even if specific bacteria exist within it or not. There will be 'speaking smart packages' planned, i.e., gift packs that can record the voice and play it back to the recipient, and talking smart packages in healthcare: pharmacies will put a radio tag under the printed instructions on medication so the sight-impaired patient can hold a gadget nearby that speaks out all the details. Smart packages have potential use in healthcare, in anti-theft tagging, as radiofrequency ID tags, in diagnostic applications, and so on [8].

5.2.5 Active Packaging (Antimicrobial Packaging with Biocidal Polymers)

Active packaging is the type of packaging that controls and reacts to events occuring inside the packaging. There are different types of active packaging systems, such as, antimicrobial packaging and the packaging that can provide a controlled gas environment for the food.

(a) **Antimicrobial Packaging with Biocidal Polymers** is used to protect contents from bacterial contamination, i.e., when a phosphonium compound is embedded in a plastic, both microorganism growth on the plastic as well as deposits of iron carbonate or iron, lead and zinc scale deposits are avoided [9]. Or in the healthcare industry, it is shown that, mono- and multi-layer medical tubing incorporating a silver-based antimicrobial can be used efficiently [10].

There is evidence showing that some of the antimicrobial packaging additives can also be very effective as indirect food additives [11, 12].

There are a variety of different geometries applied for antimicrobial packaging. In one application, co-extruded films of PE comprising a liquid-absorbent inner layer, an antimicrobial middle layer (to decrease the migration possibility of additives) and an impermeable outer layer prepared as foodstuff packaging is presented [13]. While in another, coatings with microbiocides are applied to substrates such as PP films and polyester fabrics by plasma-polymerisation and deposition [14].

Within a number of different biocides, there is also *N*-chloro-hindered amines, used as a multifunctional additive.

Antimicrobial packaging shows promise as an effective method for the inhibition of certain bacteria in foods, but barriers to their commercial implementation still exists.

(b) **Controlled/Modified Atmosphere Packaging (CAP/MAP).** This method is used for the storage, transport and packaging of perishables by surrounding the food for a controlled time period with an inert gas (carbon dioxide or other antimicrobial) inside the package [15].

(c) **High Barrier Packaging** (films with controlled extractables or gas barriers). Improvement of barrier properties for plastics are desirable for different applications, i.e., in food and beverage packaging, they are needed to slow the escape of carbon dioxide from the system or to decrease or eliminate the entrance of oxygen, in order to improve quality and extend shelf life, as well as to prevent diffusion of water vapour, air, aromas, or fuel emissions, which was accomplished traditionally by metal or silicon oxide coatings. High barrier polymers (like some polyamides, polyvinylidenechloride (PVDC) and EVOH) are usually used either as coatings or as layers in multilayer structures. Recently, nanocomposites based on nanoclays, are found to improve barrier properties with some certain advantages. In the classical case, co-extruded or laminated multilayer films with at least one barrier film and an intermediate adhesive (tie) layer can provide control of leaching of contaminants into a packaged product, or can be completely 'leach resistant' [16].

In cases where oxygen sensitive products are packaged, special layers providing a barrier specifically to the passage of oxygen are needed and used, in the form of co-extruded multilayered structures [17]. For the gas barrier to be successful, it should be able to:

a) Scavenge the oxygen that may be present in the package interiors, and/or,

b) Inhibit the migration of off-taste or off-odours to the package interior from outside.

(d) Environment Sensitive Smart Packaging and Smart Packaging with Active Integral Displays. Certain biopolymers that have the potential for applications such as controlled release devices, environmentally sensitive membranes, mimic materials and energetic applications are planned to be used for smart packaging. For this, plant or microbial biopolymers with ionic functional groups have shown promise [18].

Active integral displays are expected to be used in intelligent medical packaging applications, which would allow static as well as dynamic visual information [19].

5.3 Rubbers Used in Contact With Food and Possible Health Effects

Use of rubbers in packaging is rather rare, but there are still a number of cases where food gets in contact with rubber, hence the health issues of rubber (that are in contact with the food) should also be considered. Some of these contacts with food occur during their processing stages and/or mostly during storage of the food, i.e., because of the 'rubber seals used in the flip-top stoppers on beer bottles' or 'rubber seals used in some jar tops' as well as 'the seals at the ends of food cans' can be recalled for the latter case. The contact of rubber is more significant during the 'processing of food' stages, (i.e., on conveyor belts, hosing, seals, gaskets, mixing pedals, skirting and liners as in the case of milk industry for the latter). In such cases, the condition of the food will be important as well (such as type of food, temperature, contact time and contact area, where the last two are usually small as opposed to packaging). However, since rubber compounds are very complex chemically as they contain a number of additives (around 10-15) at all times (such as plasticisers and process aids, antidegradants, curatives and cure co-agents, accelerators, vulcanisation remnants), there are a number of potential migrating agents that exist in the system. In addition, for a rubber matrix, which is flexible with an high permeabilities, this could allow many possible additive migrations. After considering these facts, rubber, used as a food contacting material, should have more attention paid to it, than plastics.

There is the EU resolution (Council of Europe) *Resolution on Rubber*, and the Food Standards Agency, which already is funding a number of research projects to resolve the issue [20, 21].

As far as the health issues of rubber is concerned, the same problems of additives as discussed for plastics are important, which means the type and extent of accelerators, antioxidants, antiozonants, antistatic agents, extenders, fillers, (process and extender) oils, vulcanisers (mostly organic), pigments (mostly in the form of blends), plasticisers, reinforcing agents and resins, as well as monomers and oligomers left in the system should all be considered. In addition to these, various solvents (mostly aromatics like toluene and xylene) are also used in the rubber industry and their remnants can be important in some cases (i.e., solvents are used to prevent tackiness as a general application, and a number of aliphatic hydrocarbons are used for the freshening of rubber surfaces, and carbon disulfide is used in the traditional rubber cold-cure processes.). In addition, there may be other inherent factors and chemicals that may arise from the rubber itself (breakdown products), that may have health effects as well [22]. These intentional and unintentional compounds (which are called *indirect food additives* by the FDA) have the potential to migrate into the food, in the case of rubber contact with the food.

The regulations for indirect food additives are given in Title 21 of US Code of Federal Regulations, and permitted compounds are given in a positive list; considering various operational conditions (such as temperature, type of food and type of contact with the food).

5.3.1 Some Rubber Types Used in Contact with Food

There is a relatively large range of different types of rubbers that are used in different components in the food industry that can get in contact with the food. The most important of these are: natural rubber (NR; *cis*-1,4-polyisoprene), nitrile rubber (i.e., acrylonitrile-butadiene copolymer), ethylene-propylene rubber (EPR), rubbers of ethylene-propylene monomer (EPM) and EPDM, SBR, fluorocarbon rubber, silicone rubber, polybutadiene rubber (BR), polychloroprene rubber, and TPE. In addition, there is the use of rubber blends, i.e., blends of NR and N Rr with SBR [19].

Rubbers are used mainly in the seals of packaging (i.e., for this, EPDM is used commonly, while nitrile polybutadiene rubber, and its derivates such as hydrogenated nitrile rubber and acrylates, fluoro-rubbers and silicones are all used for sealing oils and fuel containers, specifically).

NR compounds have better heat ageing characteristics, and hence they are used preferably as seals and gaskets, as food can sealants and teats, and as soothers, as well as for the production of gloves. In food processing equipment, NR compounds are found in the belting and hosing products mostly (for both aqueous and fatty foodstuff), and in dairy hosing milk liners.

EPR compounds are mainly preferred in the manufacture of heat exchanger gaskets, and usually need curing (usually done by using peroxides).

Fluorocarbon rubbers are used in seals and gaskets, specifically for use at high temperatures for prolonged use, especially in contact with fatty (oily) foods.

Silicone rubbers used in the food industry are generally of 'polydimethyl vinyl siloxane' type, and are successfully applied for seals and tubing with non-sticky surfaces.

TPE have a large range of applications in the food industry, i.e., as flexible lids (specifically for styrenics), belting, gaskets and tubing (for PU). Within TPE, there are:

a) **Styrenic block copolymers** (styrene-isoprene-styrene, styrene-isobutylene-styrene, SBS),

b) **Olefinic blends** (PP with uncured EPDM, PP with cured EPDM and NR),

c) **Polyurethanes** (PU, such as the ones prepared from diphenylmethane diisocyanate, methylene diphenyl diisocyanate and polytetramethylene ether glycol/adipic acid glycols with an extender like 1,4-butanediol),

d) **Various polyesters** (based on dimethyl terephthalate, and 1,4-butanediol).

5.3.2 Issue of Monomers and Oligomers (Left) in Rubbers

Usually a small amount of monomer and oligomer are left unreacted at the end of a polymerisation reaction (such as styrene, acrylonitrile and isocyanate), concentrations of which can be critical to know - because their molecular weights are small and the matrices (rubber) are flexible, letting them migrate to the surface easily. Especially for some toxic monomers like acrylonitrile, this can be of utmost importance and permissible limits of such species must be defined and followed carefully for food contact applications (in NR, the concentration of free monomer is set at 1 mg/kg as the maximum). Similarly, oligomers that can exist in the system after completion of the polymerisation can pose similar problems.

5.3.3 Issue of Vulcanisation Agents (and Cure Products) Left in Rubbers

Vulcanisation (cure) reaction of rubbers is usually complex, and a number of different reaction products can be produced. These new (unwanted) products plus the unreacted part of the accelerators (like thiourams, certain thiazoles, sulfenamide and dithiocarbamates) are potential dangers for migration in food contact applications. Especially for rubbers contacting aqueous media, breakdown product migration is more common.

However, accelerators can also be very important to consider in all cases for rubber: certain cure accelerators (such as thiouram) can lead to the production of a (suspected) carcinogen: nitrosamines.

5.3.4 Plasticisers and Antidegradants in Rubbers

Plasticisers in rubbers are usually at high concentrations, and although concentrations of antidegradants (antioxidants and antiozonants) are much lower, they are all very critical in food contact rubber products.

The rubber industry uses various hydrocarbon additives, specifically called process oils (which act more or less as a plasticiser, used below 20 phr) or extenders (used for economy), to function as plasticisers. There are a wide range of mineral oils used as process oils. They are produced by blending crude oil distillates and there are three main grades to consider: paraffinic (with branched and linear aliphatic hydrocarbons), naphthenic (with saturated ring structures) and aromatic. These, with compounds containing sulfur, nitrogen or oxygen are the polar component of the oil. The polycyclic aromatic hydrocarbon containing process oils are classified as carcinogens, however, their use is decreasing.

5.3.5 Migration from Food-Contact Rubbers and Some Tests

Migration from a food contact rubber material into food (or into food simulants), is usually expressed in two ways, either as:

a) **Overall migration** (total extractables: mass of overall migration without considering composition of the migrant), which is FDA recommended, and/or,

b) **Specific migration** (where composition and quantity of migrant are of interest).

In the case of overall migration, the amount of 'total extractables or overall migration' obtained after repeated contacts with the rubber and food, are specifically sought. This method states that 'the food contact surface of rubber should not yield to more than 129 mg/cm^2 of total migrants to extract (into distilled water as the simulant) for the first 7 hours of extraction, and a maximum migration of 6.45 mg/cm^2 during the next 2 hours'; or for fatty foods, 'the food contact surface of rubber, should not yield more than 1129 mg/cm^2 of total migrants to extract (into *n*-hexane as the simulant) for the first 7 hours of extraction, and a maximum migration of 25.8 mg/cm^2 during the next 2 hours' [19].

And for both of these tests; 'contact times', 'temperature' and 'the size of contact area' are all very important, where the first of these contact times are considerably small in the case of rubbers specifically (in the order of minutes or even shorter), which is very different if compared to plastics (weeks or more).

There are also the 'classical **German BgVV tests**', where there are three categories which are dependent on the time of contact of rubber with food:

a) If contact is for more than 24 hours: Category 1 (i.e., storage containers, linings, seals in cans),

b) If contact is for up to 24 hours/but more than 10 minutes: Category 2 (i.e., bottle stoppers),

c) If contact is for less than 10 minutes: Category 3 (i.e., milk liners, milking machine tubes and accessories), and

d) Other cases: special category (i.e., toys, balloons, baby teats, breast caps, teething rings, and so on).

In fact, a number of different rubbers (such as NR, silicone, SBR) have been commonly used as teats for baby feeding bottles, where these migrations, in particular, the level of nitrosamines are of utmost importance. Several other references can be cited on the topic [23-25].

There is really not much of published data available on the migration of chemicals from rubber into food (or food simulants). Within these, there is the Smithers Rapra study (and report) worth mentioning first, accomplished by use of a series of specifically compounded different types of rubber samples, which shows migration of a number of amines, N-nitrosamines and aldehydes (mainly formaldehyde), from the prepared rubber compounds of NR, EPDM rubber, fluorocarbon rubber and silicone rubber. It concludes that, specific migratory components are from decomposition of reaction products of the vulcanisation system and that rubber samples cured with sulfur (and all accelerators containing sulfur) showed characteristically high levels of extracted N-nitrosamines and aromatic amines [19]. In addition to this, the preparation of a permitted list of additives (positive list) for food-contacting rubber compounds is aimed at, in an another study [26]. In the rather classical study undertaken in Poland in 1981, migration of rather high amounts of metal ions (arsenic, barium and lead) were found in 30%, and migration of the known carcinogen phenyl-naphthylamine were found in 15% of the rubber samples tested [27]. Substantial migrations of dimethylamine (from stoppers, spatulas and teats) into water and hydrochloric acid, were noted, in another study done in Japan [28].

References

1. J.H. Petersen, X.T. Drier and B. Fabech, *Food Additives and Contaminants*, 2005, **22**, 10, 938.

2. M.W. Kid, *Asian Journal of Chemistry*, 2005, **17**, 1, 40.

3. A. Feigenbaum, P. Dole, S. Aucejo, D. Dainelli, C. De La Cruz Garcia, T. Hankemeier, Y. N'gono, C.D. Papaspyrides, P. Paseiro, S. Pastorelli, S. Pavlidou, P.Y. Pennarun, P. Sailor, L. Vidal, O. Vitrac and Y. Voulzatis, *Food Additives and Contaminants*, 2005, **22**, 10, 956.

4. E. Weise, *USA Today*, 2005, 8A.

5. R. Czarnecki, *Ink Maker*, 2005, 83, 2, 29.

6. *SpecialChem Innovations and Solutions through Polymer Additives & Colors*, SpecialChem, www.specialchem4polymers.com

7. *Comprehensive Chemical Legislation Database (CHEMLEX)* on CD-ROM, UN/ECE, Geneva, Switzerland, 1998.

8. *The New Omnexus, Innovations and Solutions Through Plastics and Elastomers*, SpecialChem-omnexus, http://omnexus.com/

9. C.R. Jones and R. Diaz, inventors; Rhodia Consumer Specialties Limited, assignee; WO 5079578A2, 2005.

10. I. Moore, J. Godinho, C.Y. Lew, G.M. McNally and W.R. Murphy in *Proceedings of the SPE ANTEC Conference*, Boston, MA, USA, 2005, p.3130.

11. K. Cooksey, *Food Additives and Contaminants*, 2005, 22, 10, 980.

12. J. Yan, inventor; Avantec Vascular Corporation, assignee; WO 5073091A2, 2005.

13. I. Lee and K. Hausmann, inventors; E.I. DuPont De Nemours and Company, assignee; WO 5113236A2, 2005.

14. A.J. Goodwin, S.R. Leadley, L. O'Neill, J.P. Duffield, M.T. McKechnie and S. Pugh, inventors; Dow Corning Ireland Limited; assignee, WO 5110626A2, 2005.

15. M. Koyama, M. Kogure, Y. Oda, Y. Maruhashi, T. Goryoda, S. Mukuno, inventors; Toyo Seikan Kaisha, Ltd., assignee; US 6964796B1, 2005.

16. C.E. Altman and D.J. Gibboni, inventors; Honeywell International, assignee; US0252818A1, 2005.

17. S. Dick, I. Kraemer, G. Goldhan, N. Rodler, T. Hubensteiner, C. Stramm, K. Rieblinger, inventors; Süd-Chemie AG and Fraunhofer-Gesellschaft zur Foerderung der Angewandten Forschung eV, assignees; WO 5108063A1, 2005.

18. V. Finkenstadt and J.L. Willett, *Macromolecular Symposia*, 2005, 227, 367.

19. G.L. Bradley, *IP.com Journal*, 2005, 5(9B), 18.

20. J.A. Sidwell and M.J. Forrest, *Rubbers in Contact with Food*, 2000, Rapra Review Report No.119, Volume 10, No.11, Rapra Technology, Shrewsbury, Shropshire, UK.

21. M.J. Forrest, *Food Contact Rubbers 2 – Products, Migration and Regulation,* 2006, Rapra Review Report No.182, Volume 16, No.2, Rapra Technology, Shrewsbury, Shropshire, UK.

22. J.S. Taylor and Y.H. Leow, *Rubber Chemistry and Technology,* 2000, **73**, 3, 427.

23. K. Mizuishi, H. Takeuchi, H. Yamanobe and Y. Watanabe, *Annual Report of Tokyo Metropolitan Research Laboratory of Public Health,* 1986, **37**, 145.

24. J.B.H. Lierop in *Food Policy Trends in Europe: Nutrition, Technology, Analysis and Safety,* Eds., H. Deelstra, M. Fondu, W. Ooghe and R. Van Havere, Ellis Horwood, UK, 1991, p.215.

25. N.P. Sen, P.A. Baddon and S.W. Seaman, *Food Chemistry,* 1993, **47**, 4, 387.

26. H. Jarrijon, *Revue Generale des Caoutchoucs et Plastics,* 1993, **70**, 725, 67.

27. H. Mazur, L. Lewandowska and A. Stelmach, *Roczniki Panstwowego Zakladu Higieny,* 1981, **32**, 97.

28. T. Baba, M. Saito, Y. Fukui, S. Taniguchi and Y. Mizunoya, *Journal of the Food Hygienic Society of Japan,* 1980, **21**, 1, 32.

Some Additional Related Literature

1. R.B. Simpson, *Rubber Pocket Book,* Rapra Technology, Shawbury, Shrewsbury, UK, 2002.

2. M. Ash and I. Ash, *Handbook of Plastic and Rubber Additives,* 2nd Edition, Synapse Information Sources, Endicott, NY, USA, 2005.

3. B.G. Willoughby, *Air Monitoring in the Rubber and Plastics Industries,* Rapra Technology, Shawbury, Shrewsbury, UK, 2003.

4. J.M. Vergnaud and I.D. Rosca, *Assessing Food Safety of Polymeric Packaging,* Rapra Technology, Shawbury, Shrewsbury, UK, 2006.

5. D.J. Knight and L.A. Creighton, *Regulation of Food Packaging in Europe and the USA,* Rapra Review Reports, 2003, **15**, 5, 173.

6. R.H.D. Beswick and D.J. Dunn, *Plastics in Packaging-Western Europe and North America,* Rapra Technology, Shawbury, Shrewsbury, UK, 2002.

7. *Plastics in Food Packaging: Properties, Design and Fabrication*, Ed., E.W. Brown, Marcel Dekker, New York, NY, USA, 1992.

8. B. Piotrowska in *Toxins in Food*, Ed, W.M. Dabrowski and Z.E. Sikorski, CHIPS Publications, Weimar, TX, USA, 2005, p.313.

9. D.R. Jenke, J.M. Jene, M. Poss, J. Story, T. Tsilipetros, A. Odufu and W. Terbush, *International Journal of Pharmaceutics*, 2005, **297**, 1-2, 120.

10. S. Ojha Satyajeet and V. Patil Nilesh in *Proceedings of an International Conference on Trends in Polymers & Textiles*, New Delhi, India, 2005, p.272

11. T.C. O'Riordan, H. Voraberger, J.P. Kerry and D.B Papkovsky, *Analytica Chimica Acta*, 2005, **530**, 1, 135.

6 Plastics Use in Healthcare and Their Possible Health Effects

6.1 Plastics in Biomedical and Healthcare Applications

Biomedical and healthcare applications of plastics are increasing and becoming more and more demanding, varying from sophisticated ones (such as sutures, wound-closures, drug delivery systems, orthopaedic products, angioplasty balloons, stents, various implants, and even as drugs and therapeutics by themselves; that will be referred to as low volume/high value specialty products), to some consumables (such as transfusion-infusion-injection supplies, disposables like gloves, masks, gowns, wraps, device packaging and other coverings; all these will be referred to as high volume/commodity products [1]). Although the first group of these use a series of very special and expensive, production-limited polymers; the second group depends mostly on commodity type plastics, which are more economical and are produced on a much larger scale.

Biocompatibility is a very critical, basic safety requirement for medical polymers in medical device applications. For this, it is essential that:

a) The final medical device must be biologically safe for its intended use, must be biocompatible (should comply with related standards, such as **ISO 10993 [2], EN 30993 [3], FDA 21 CFR 58**), and it should not produce any adverse effects,

b) The sterilisation processes and the intended shelf lives must be correct.

In these, production of any adverse effects can be translated as 'the possible migration of any toxins should not exist', while the last part regarding sterilisation, 'no toxic degradation products should be allowed' can be understood. For these, the basic plastic material will be of importance.

Plastics have a basic role in the medical device and packaging industries, mostly for the *low volume/high value specialty products*, because:

a) They are often chemically inert and resistant to harsh disinfectants, corrosion and sterilisation conditions.

b) They can be processed into the final required shapes and with precise configurations more easily and affordably (that would be more difficult or even impossible with other competing materials, such as glass and metals).

c)   They can be made to have proper mechanical properties (i.e., strength and impact resistance – *low breakage values, with proper flexibilities required*).

d)   They can provide much lower specific weights and higher specific strengths than their traditional counterparts (metals such as aluminium, steel and titanium as well as glass and ceramics), and mainly for the *high volume/commodity products* group.

e)   The ones used on a large scale have a relatively low cost and they certainly are more practical to use (hence they can often be disposed of and therefore need not be sterilised for re-use).

With these unique advantages it may even be claimed that, 'polymers have helped to reform the health system' of most of the world quickly and to a large extent, with a number of demographic factors, particularly with the increase of ageing population in the industrialised part of it and hence acceleration in the demand for healthcare services and medical devices. In fact, around 3.5 - 4 million tonnes of plastics is being consumed per year by the global medical technology industry, and with an annual growth rate of 6-10%, this is already one of the most attractive fields of application for plastics [1, 4, 5].

6.1.1 'Commodity' and 'Specialty' Medical Plastics

Within the *commodity-high volume plastics* used in the medical market, there are polyethylene (PE), polyvinylchloride (PVC), polystyrene (PS) and polypropylene (PP), in addition to silicones, polycarbonate (PC), polyurethane (PU), polytetrafluorethylene (PTFE) and acrylics. These are high volume polymers, and if only the market for polymeric medical packaging materials (both for medical devices and pharmaceuticals) is considered, it is many times bigger than the specialty medical plastics market for medical devices. In the polymeric medical packaging materials, polyesters are used in addition to the polymers mentioned previously as major commodity polymers, all of which constitute 5% of the overall US packaging material market (equating to US $4 billion). Disposable medical products (such as syringes, tubing, bags, gloves and sheets) produced from commodity plastics account for nearly 80% of the polymer consumption in this industry segment [6].

Specialty plastics such as PC, acrylonitrile-butadiene-styrene (ABS), PU, polyamides (PA), thermoplastic elastomers, polysulfone and polyetheretherketone (PEEK) are being used in a number of special applications in medical devices, whenever high performance is required. There are also a number of polymer blends used in this category (i.e., PC/ABS and PC/polyester), if property envelopes are required to be further improved.

6.1.1.1 PVC as a Medical Plastic

Around a quarter of all plastic medical products are made of the commodity plastic PVC (i.e., blood bags/tubing, catheters, dialysis equipment, mouthpieces and masks, oxygen delivery equipment, labware, and medical packaging), because it has low cost, is easily adaptable to a number of applications, and it is easy to process.

In medical grade PVC, *heat stabilisers* (mainly various calcium-zinc formulations) are used for its processing, storage or autoclaving. Although barium-zinc additives are more effective as heat stabilisers for PVC, in several countries they are restricted for medical applications.

As for the *plasticisers* for medical grade PVC, some compounders have developed proprietary plasticiser formulations for medical PVC that are alternatives to the common di-2-ethyl hexyl phthalate (DEHP) (for which some health concerns are involved, as outlined in Chapter 3). The Food and Drug Administration (FDA) has notified hospitals, physicians, and others involved to reduce use of DEHP-plasticised medical devices in certain patient populations, particularly 'neonatal males'. Although the FDA's guidance recognises that the vast majority of patients have a minor risk from exposure to DEHP, it recommends the use of replacement materials and voluntary labelling on PVC tubing and other devices that contain DEHP. It has already been shown that PVC medical tubes can migrate substantial amounts of plasticiser to patients, if it contains certain plasticisers (in one study, analysis of used PVC medical tubes for residual dioctyladipate showed that 64-67% had migrated to the patient, and for dioctyl adipate (DOA), 70-100% was retained in the tube [7]). DOA is known as a less potentially toxic (or regulated) plasticiser.

There are concerns for the use of PVC for drug delivery applications, and it is shown that there may be a PVC-drug interaction, causing inadequacy of the method [1].

There are also cases where the surface structure of PVC is altered using UV irradiation resulting in a significant decrease in the migration of phthalate plasticiser [8].

6.1.1.2 Polypropylene as a Medical Plastic

PP has inherent good barrier properties and high clarity, in addition to proper radiation resistance. Properties which made PP one of the best candidates in medical devices and packaging applications (parenteral nutrition and dialysis films, blister packaging and flexible pouches, syringes, tubing, hospital disposables, test tubes, beakers and pipettes). Medical grade PP is used mostly as blown, cast films and also as coextruded layered structures.

6.1.1.3 PS and its Copolymers as a Medical Plastic

PS, both crystalline and high impact (HIPS) types, has low cost, low density, good clarity and high dimensional stabilities, in addition to the ease of adaptability to radiation sterilisation [9]. During sterilisation with ionising radiation, polymer and additive degradation may result in new toxic products - PS is safe in this aspect. *Crystalline PS* is mostly used as a commodity/high volume medical plastic (as labware such as Petri dishes and tissue culture trays), while HIPS is used (preferably) as a specialty plastic (used mainly in thermoformed products, such as catheter trays, heart pump trays and epidural trays), where it is competitive with PVC, PP and acrylics. Both crystal PS and HIPS are used in respiratory care equipment, syringe hubs and suction canisters. In labware and packaging for kits and trays, PS is competitive with PVC, PP and acrylics. There are also studies done by plasma to improve the cell adhesion on PS surfaces [10]. Expandable PS foams are used preferably for blood and organ transport between 2 °C and 8 °C.

The main medical plastics of styrene copolymers are two types: ABS and styrene acrylonitrile (SAN).

ABS is rigid and tough, with excellent impact, chemical and radiation resistances, and it complies with ethylene oxide sterilisation. ABS is used in surgical instruments, roller clamps, piercing pins, instrument housings, diagnostic test kits, and hearing aid housings.

SAN is relatively more economical, and has a high clarity, chemical resistance, ease of processability, with high hardness and stiffnesses, and has an attractive surface gloss. SAN is used preferably in urine meters, flat plate dialysers, disposable diagnostic kits and fluid handling devices.

6.1.1.4 PE as a Medical Plastic

PE is mostly used as a commodity/high volume medical plastic: high density PE is a common polymer used for medical tubings and pharmaceutical closures, while LDPE is mostly used in sterile blister packs for drug packaging.

Another version of PE, the ultra-high molecular weight polyethylene is used as a specialty plastic (in artificial hip, knee and shoulder joints), mainly due to its exceptional impact strength, low wear, good stress-crack resistance, durability, and excellent energy absorption characteristics.

6.1.1.5. PC as a Medical Plastic

Medical grade polymers of PC are used specifically for haemodialysis filter membranes, surgical instrument handles, and the housings of oxygenators (all are characteristic devices that enrich blood in oxygen and remove carbon dioxide, i.e., during open heart surgery), in addition to use in needle-free injection systems, in perfusion equipment, blood centrifuge bowls and stopcocks. Corrective lenses for eyes are often made of PC. PC is exceptionally tough, highly transparent, strong and rigid, having exceptional resistance to steam sterilisation. PC medical products can be easily sterilised with ethylene oxide, gamma radiation, electron beam radiation and steam autoclaves. PC plastics are the fastest growing material group in the medical device and packaging sector, mainly because of their highly favourable combinations of 'cost and performance' factors. In fact, PC are not the most widely used plastics in medical products, like PVC, PE, PP and PS. But PC offers high levels of heat and radiation resistance that neither PVC nor other commonly used medical plastics can offer. In addition, PC economy is much more favourable than that of other high performance engineering plastics used in medicine, such as polysulfone and PEEK. In addition to the standard PC grade resins, there are also improved special grades of PC possible (i.e., high temperature grades that can be autoclaved at temperatures as high as 134 °C more than once, radiation resistant grades (introduced to avoid yellowing), and lipid resistant grades (introduced to avoid stress cracking of the material due to attack of lipids/fats and oils, which may exist in intravenous solutions such as lipid-based emulsions). PC are generally biocompatible, and certain commercial grades of PC are already certified to meet specified biocompatibility standards, such as ISO 10993-1 [2]. There are also studies done to further modify the surface properties of biocompatible PC membranes to encapsulate pancreatic islets [11]. Blends of PC with other polymers (such as ABS and polyester), are commonly prepared to combine the strength and rigidity of PC with the high-flow properties of ABS, or the chemical resistance of polyesters, and they are mainly used in the housings of many medical instruments [12]. One of the smallest portable public access defibrillator (PAD) made of a thermoplastic elastomer and an ABS/PC blend is available in the USA and in Europe. For needle free drug delivery injection systems (including intramuscular injections, subcutaneous injections and intradermal injections) PC, and for the needle free injector, a blend of ABS/PC have been selected.

Emerging applications of PC include 'inhalers' for the consumer market, ophthalmic products, orthopaedic materials, and 'lab-on-a-chip' type of devices to determine blood chemistry/or to analyse proteins.

6.1.1.6 Thermoplastic Polyurethanes as Medical Plastics

Thermoplastic polyurethanes (TPU) have excellent clarities, high tensile and tear strengths, good grades of chemical and abrasion resistances, have excellent hydrolytic stabilities and smooth surfaces that show perfect fungi and microorganism resistances.

Medical grade TPU are used mainly in transdermal patches, medical tubing, oxygen masks, catheters, drug delivery devices, IV connectors and cuffs. They are also used for bone repair (orthopaedic) applications and hip implants. Gloves made from PU are claimed to be more durable and safer than latex types. PU fibre and polyester-based wound dressings are very helpful as first aid materials, because they are an instant pressure-dressing impregnated with coagulant and antibiotic [13]. There is also a low pressure wound dressing developed by use of multi-density PU foam to speed up healing of ulcers, burns and wounds. In all of these applications, the danger of migration of monomers and additives left in the system should be remembered and accounted for.

6.1.1.7 Polyamides as Medical Plastics

Medical grade PA are not usually affected by body fluids, and they do not cause any skin or tissue inflammation, and some of its grades have been already approved for blood contact, which allows them to be used in transfusion equipment (in connectors, adapters and stopcocks). A Nylon 12 catheter has a specifically low degree of interaction with body tissues, hence it can be inserted with minimum pain for the patient. PA are known for their high flexibilities, hardness and toughnesses, and resistances to warping.

6.1.1.8 Some Engineering Polymers as Medical Plastics (USP Class VI type)

Silicone elastomers, with their low surface energies, hydrophobicities, as well as chemical and thermal stabilities, are widely used throughout the medical device industry (for component fabrication) and in pharmaceutical applications (tubings for fluid-handling), as a hybrid material (in catheter applications), and as implants.

Polysulfones are high temperature resistant and can be sterilised by all known methods effectively and easily. They are often used for sterilisable vacuum-formed medical equipment.

Polyethersulfone (or polyarylsulfone) can withstand exposure to cold sterilants, disinfectants and germicides, and its applications are mainly in surgical tools.

Polyphenylsulfone (or polyarylethersulfone), can be steam sterilised repeatedly, it also resists acids and bases. Typical medical applications of the material include sterilisation trays, surgical instruments and fluid handling equipment.

Polyphenylene sulfide (PPS) is known for its dimensional stability, toughness and rigidity. PPS can be sterilised repeatedly and can be exposed to strong disinfectants without damage. PPS has been used to replace metal in the precision mechanical elements of drug delivery systems.

Liquid crystal polymers (LCP) provide excellent strength, stiffness, dimensional stability and creep resistance. They are not damaged by the high temperatures commonly used for sterilisation. LCP are often used to replace metals in intricate drug delivery system components, and are also found in devices used for minimally invasive surgery. **Thermoplastic Elastomers (TPE)** have the softness and flexibility of rubber, but can be processed by conventional thermal techniques and equipment. TPE can resist extremes of temperature, (i.e., between 40 °C and 121 °C), can resist common solvents, oils, dilute acids and bases. TPE are commonly used as tubing, stoppers, transdermal patches and intravenous bags in the medical industry.

Thermoplastic polyesters include polybutylene terephthalate (PBT), which possesses good slip and wear properties, resists chemicals and hydrolysis, and has excellent dimensional stability. Mechanical parts in drug delivery and filter systems are preferably made of PBT.

Acetals (or polyoxymethylene) have similar properties to PBT.

Polyetheretherketone is a high-performance engineering resin that is used in highly demanding medical applications. PEEK has a continuous use temperature of 260 °C and it can be exposed continuously and periodically to hot water and steam without loss of properties. Excellent dimensional stability and high impact and tensile strengths are the characteristics of PEEK. In the medical field, PEEK is used for implantable medical devices, such as a spinal cage (in place of titanium) and finger joints, and as tubing in minimally invasive surgery (such as stent placements in heart arteries), as well as direct catheters, drug delivery devices, surgical instruments and endoscopes. **Acrylics** - which include polymethyl methacrylate (PMMA) are in general, highly transparent, strong, chemically resistant and compatible with biological tissues. In the medical field, acrylics are used in tubing connectors, cuvettes, speculums and other devices that require high clarity. Acrylics compete for some of the same markets as PC.

Polyether-imide (PEI) is an amorphous thermoplastic, with an excellent balance of physical properties and dimensional stabilities. PEI can be used with the full spectrum of sterilisation methods. Surgical probes that are subjected to repeated cleaning and sterilisation are their typical preferred application as a medical plastic.

Fluoropolymers (such as PTFE), have low friction coefficients and high tensile strengths (plus good chemical resistances) are the best common choices for catheters.

Silicones are also commonly used for catheters and fluid drainage devices. Their high degree of biocompatibility is the main reason that silicones are chosen over lower priced tubing materials. Silicone polymers are also used as standard materials for artificial finger joints [14].

6.2 Fibre Reinforced Plastics as Medical Materials

Reinforced plastics are also introduced in some special applications as medical plastic materials, i.e., carbon fibre reinforced plastics are being used in Germany to replace metal in certain medical instruments such as a special puncture needles. This system has the advantage of not affecting magnetic resonance imaging equipment and to be chemically resistant to a number of aggressive media. The puncture needle is a multi-function needle containing three hollow optical fibres.

6.3 Direct Use of Synthetic Polymers as Drugs and Therapeutic Agents

Synthetic polymers are of increasing interest as therapeutic agents and drugs as well, owing to their enhanced pharmacokinetic profiles relative to small molecule drugs.

These so-called 'polymer therapeutics' are believed to be likely to form the 'next generation' of medicines, with their improved efficacy against a wide spectrum of diseases [15].

6.4 Dental Resin Composites

A variety of resin composite systems are used currently in dental restorations, because of their strength, rapid in-cavity polymerisation possibilities, good levels of adhesion to enamel and dentin, in addition to aesthetic appearance. They are polymerised (crosslinked) in place, by UV and in their initial original mixture there are monomers, initiators and co-initiators, additives, fillers, as well as a number of by-products; most of which are soluble in water to a certain extent. It is shown that, it is possible to have subsequent release of these constituents, including the monomers (which are: triethylene glycol dimethacrylate (TEGDMA), hydroxy ethyl methacrylate (HEMA), urethane dimethacrylate, bisphenol A glycidyl dimethacrylate (BISGMA), bisphenol A dimethacrylate and bisphenol A), both from dental resin composite systems and resin-modified glass ionomer cement composites [16, 17]. Release of monomers are different for different monomers and conditions, but could be very high (as high as 500 µmol/m^2 surface area), mostly during the first day [18]. Some of these monomers (HEMA, BISGMA) are known to have been associated with inflammatory pulp responses, contact dermatitis and allergic reactions, while some others (TEGDMA) are known to cause alteration of the growth and lipid metabolism of oral epithelial cells, and bisphenol A, has been associated with oestrogenic responses in human cells [19-21].

6.5 Use of Polymers in Dialysis

Dialysis is a vital and expensive application and there are a number of polymers used in dialysis. Since all of the components of the extracorporeal blood circuit are in contact with blood under dynamic conditions, the use and selection of polymers are very critical. The following polymers are used as dialysis polymers: **PVC** (in blood tubing), PC and PP (in dialyser housings), PU (as a potting material for capillary membranes at both ends of the dialyser to separate blood compartment from dialysis fluids), silicone rings (to prevent any blood leakage), membrane polymers (regerated cellulose, modified cellulose, PS, PMMA, polyacrylonitrile, PA, EVOH). All of these polymers, in addition to being extremely pure, must be blood compatible [22].

6.6 Ophthalmic, Prostheses and Other Applications of Medical Polymers

Soft hydrogels made of HEMA are a well-known common example of medical plastics in contact lenses. They can absorb water and can transmit oxygen, and with these properties, they are replacing previous acrylic hard contact lenses. Replacement prostheses are successfully being applied in place of a number of organs in the human body. Both in ophthalmic and prostheses applications, the durabilities as well as biocompatibilities of the material are too critical to consider, because the parts are supposed to function under aggressive conditions for a long time. In addition, antioxidants that have to be used in these systems to give durability, should be very carefully selected, because they can slowly and easily leach into body fluids in both cases. The immobilisation studies of antioxidants and stabilisers to the polymer backbone is still not completed, after which the danger of leaching most probably will be over.

Plastic sutures (and supporting meshes) are, depending on the location of their application on or in the body, can be permanent or biodegradable in time with body fluids, and so additive and biocompatibility issues should be considered critically and carefully.

Bone cements are vital to use in hip joint operations as a glue. It is polymerised *in situ*, from the monomer, when catalysts (usually benzoyl peroxide) and additives are all mixed together at the place of application (in the body, between the bones to be glued). Besides the slight increase of temperature locally (due to the exothermic polymerisation reaction), which is a discomfort certainly, there will be danger from remnants of monomer that are left, after the reaction, at the site of application in the body [23].

References

1. B.J. Lambert, F.W. Tang and W.J. Rogers, *Polymers in Medical Applications*, 2001, Rapra Review Report No.127, Volume 11, No.7.

2. ISO 10993, *Biographical Evaluation of Medical Devices*, 2003.

3. EN 30993-1, *Biological Evaluation of Medical Devices-Guidance on Selection of Tests*, 1994.

4. *The New Omnexus, Innovations and solutions through plastics and elastomers*, SpecialChem-omnexus, http://omnexus.com/

5. *High Performance Plastics*, 1999, **8**, 3.

6. R. Bhardwaj, *Popular Plastics & Packaging*, 2005, **50**, 2, 84.

7. Q. Wang and B.K. Storm, *Polymer Testing*, 2005, **24**, 3, 290.

8. R. Ito, F. Seshimo, Y. Haishima, C. Hasegawa, K Isama, T. Yagami, K. Nakahashi, H. Yamazaki, K. Inoue, Y. Yoshimura, K. Saito, T. Tsuchiya and H. Nakazawa, *International Journal of Pharmaceutics*, 2005, **303**, 1, 104.

9. F. Welle, *Pharmazeutische Industrie*, 2005, **67**, 8, 970.

10. S.A. Mitchell, M.R. Davidson, N. Emmison and R.H. Bradley, *Surface Science*, 2004, **561**, 1, 110.

11. L. Kessler, G. Legeay, A. Coudreuse, P. Bertrand, C. Poleunus, E.X. Vanden, K. Mandes, P. Marchetti, M. Pinget and A. Belcourt, *Journal of Biomaterials Science*, Polymer Edition 2003, **14**, 10, 1135.

12. *Polycarbonates See Healthy Growth in Medical Applications*, OmnexusTrendReports, http://omnexus.com/resources/articles/article.aspx?id=11673

13. *Urethanes Technology*, 2005, **22**, 2, 36.

14. *Polymers Help Rebuild Damaged Body Joints*, Omnexus Omnexus Trend Reports, <ttp://omnexus.com>

15. B. Twaites, C. de las Heras Alarcón and C. Alexander, *Journal of Materials Chemistry*, 2005, **15**, 4, 441.

16. S. Lee, H. Huang, C. Lin and Y. Shih, *Journal of Oral Rehabilitation*, 1998, **25**, 8, 575.

17. W. Spahl, H. Budzikiewicz and W. Geurtsen, *Journal of Dentistry,* 1998, **26**, 2, 137.

18. S.A. Mazzaoui, M.F. Burrow, M.J. Tyas, F.R. Rooney and R.J. Capon, *Journal of Biomedical Materials,* 2002, **63**, 3, 299.

19. S. Boillaguet, M. Virgillito, J. Wataha, B. Ciucchi and J. Holz, *Journal of Oral Rehabilitation,* 1998, **25**, 1, 451.

20. K. Soderholm and A. Mariotti, *Journal of American Dentists Association,* 1999, **130**, 2, 201.

21. C. Hansel, G. Leyhausen, U. Mai and W. Geurtsen, *Journal of Dental Research,* 1998, **77**, 1, 60.

22. V. Jörg, *Medical Device Technology,* 2001, **12**, 1, 18.

23. *Polymers in Medicine: Biomedical and Pharmaceutical Applications,* Eds., R.M. Ottenbrite and E.Chiellini, Technomic Publishing Company Inc., Lancaster, PA, USA, 1992.

7 Plastics and Rubbers Applications in Construction and Their Possible Health Effects

7.1 Introduction

The building and construction sector is the second largest user of plastics (over 20% of the total consumed, which amounts to more than 6 M tonnes in the EU). Most of these plastics are commodity (the biggest share belongs to polyvinylchloride (PVC), followed by polystyrene (PS), polyolefins, polyurethanes (PU), epoxies, and others (i.e., reinforced plastics composites). They are mainly used in structural-load bearing, as well as for cosmetic, protective, rehabilitation repair, retrofitting, and insulation applications.

The application of rubber, on the other hand, is much less and is used for specific cases only (i.e., in window profiles, and in roofing), in the construction sector.

Some of these plastics and rubbery materials can cause some health hazards to humans, mostly due to the additives that they contain. These hazards are in addition to other possible sources of toxins indoors. In fact, the environment of modern society, both indoors and outdoors, can (unfortunately) full of a number of toxic chemicals. And specifically in the case of indoors, the concentration of toxics can be even higher and more critical than their counterparts in the outdoors, because there is a closed environment involved with the inside. According to a study by US Federal Environmental Protection Agency (EPA), *'indoor air is often a greater source of exposure to hazardous chemicals than is outdoor exposure'*.

Most of the toxics may already exist naturally indoors, such as radon, or they may come from the materials of construction (such as plastics and rubbery flooring materials, wallpaper, treated wooden structures, various treated furniture, paints and so on), as well as from common household products such as cleaning agents.

In this chapter, possible chemical toxics in the indoors atmosphere as related to the existence of some plastics, and treated plastics, as well as rubber construction materials (which may also be referred to as indoor chemical contaminants) will only be outlined, with some additional general brief information about radon and its effects. Hence, biological contamination will be totally ignored although in over 40% of the cases, there is biological contamination involved due to bacteria, fungi, moulds, pollens, arachnids, insects, and so on, which can lead to various allergens and health problems as well.

It should be emphasised and noted here, that the use of plastics should not automatically mean that its use always will result in adverse health effects. In fact, there are a number of different plastics being used safely in many critical health related applications such as food packaging materials and in the health sector directly, as blood bags and dialysis equipment tubing, which shows that in most cases it is not the plastic itself but the additives and other foreign chemicals that are added to the plastic which can pose health hazards and should therefore be considered carefully, as discussed in previous chapters.

Thus, in some cases, the same plastic material can be very safe or very hazardous, depending on the ingredients used to produce it.

Exposure of humans to all types of toxics existing in the indoors atmosphere is usually done unintentionally, but it can lead to very serious health issues for humans. Effects of these toxic compounds on health are usually neglected and, in fact, they are one of the least known issues in our living space. However, as given in the following examples, they should be considered more seriously because they can be very important and even vital, and, they may be the reason for a number of (hidden) health problems, beginning from a long-lasting allergy or asthma or even a long-lasting chronic headache, to more serious issues, like cancer.

To begin with, firstly, general aspects of indoor air quality (IAQ) and sick building syndrome will be discussed, which then will be followed by the other possible 'toxic' issues.

7.2 Indoor Air Quality and Sick Building Syndrome

IAQ and sick building syndrome are also known as building related illnesses. We usually spend most of our time indoors where chemical concentrations can be significantly higher than outdoors, and hence air quality in homes and in offices is a matter of ever-increasing concern. Volatile organic chemicals (VOC) emitted by building materials, furnishings, cleaning products, carpets and other materials found or used indoors as well as occupant activities, can accumulate to detectable (and sometimes to harmful) concentrations, hence they should be considered seriously in most cases. According to a study by the US EPA, 'indoor air is often a greater source of exposure to hazardous chemicals than is outdoor exposure'.

In fact, the EPA has listed both IAQ and sick building syndrome as 'one of the top five environmental problems'.

Adverse health effects with increased VOC concentrations can begin first with eyes and respiratory irritation (including asthma), then become associated with irritability, inability to concentrate and sleeplessness later, and can end up with various disorders in health and even with cancer. In a report [1], it is shown that 7-10% of the population suffers ill health, usually as a direct result of poor IAQ.

Sick building syndrome (or as it is sometimes called Sick House Syndrome) is directly connected to IAQ and it is simply due to the 'poor indoor air quality'.

7.2.1 What is Sick Building Syndrome?

Sick building syndrome can be a serious air quality problem at home, as well as in work places. An area can be described as 'sick' mainly because people develop symptoms of illness such as headache, watery eyes, nausea, throat irritation, skin disorders and fatigue when spending considerable time indoors where there is a buildup of air pollutants from household products, building materials, formaldehyde and/or respirable particles, and there is no precise definition of sick building syndrome. Signs of a sick house usually include a musty, stuffy smell and other odours at first. Moisture build-up indoors plays a large part in sick building syndrome since high humidity increases the emissions of odours and chemicals such as formaldehyde, and it promotes the spread of mildew which can aggravate or cause allergies.

In general, sick building syndrome sufferers report relief of their symptoms once the source of exposure is cut off. Several years ago, one of the major source of sick building syndrome was found to be new carpeting in houses, which can release toxic vapours for a long time after it is placed, and this has been regulated recently to enforce certain standards for a carpet's emissions. There is also considerable interest in other synthetics, such as vinylics with phthalates, as source of some sick building syndrome problems, which is still under investigation.

As regards the 'diagnosis' of a sick building, the rule of thumb is as follows: when at least 20% of building occupants 'complain of the same medical symptoms from an unknown cause for at least two weeks', the building can be suspected of being 'sick'.

One should be aware that the name sick building syndrome is rather a misnomer in as much as the syndrome can only be diagnosed by assessing the health of the building occupants, not by an examination of the building itself. A recent study suggests that the building itself, in the sick building case, should be considered as 'dysfunctional'.

The oil crises during 1973-1974 and 1980-1981 certainly agitated the development of super-tight, highly insulated houses in an effort to make homes more energy efficient. As building enclosures become tighter to reduce the exchange of air between indoor and outdoor environments in building technology, the less effective is the dilution of pollutants in the indoor space. Although there are certain claims for the existence of no correlation between tight houses and health problems, still some tightly built, well-insulated and vapour-sealed houses are known to develop signs of a sick house especially during winter months, in moderate and cold climates. The cure for this is proper ventilation, because cool air holds less moisture and replaces the air with moisture and contaminants. In warm, humid climates sick building syndrome can occur during summer months when outside air is very moist. Infiltration and ventilation, which bring humid outside air in, may increase mildew and other moisture related problems when air conditioning does not provide sufficient dehumidification.

In most cases, the ideal relative humidity range should be between 37% and 55%.

It is estimated that, some 800,000 to 1,200,000 buildings in the US are 'diagnosed' as sick. In 1987, the Polk County Courthouse, in Barstow, FL, USA, constructed for $37 M, had to be demolished and re-built again with an additional cost of $26 M in order to 'cure' the building of its 'sickness', which had necessitated the relocation of over 600 occupants of the courthouse due to their claims of sick-building symptoms. There are a number of court rulings known, mostly in US, regarding sick building syndrome cases, i.e., cases of long-term disability claims and court orders for employers claiming total disability as a result of sick building syndrome [2, 3].

7.2.2 Possible Sources of IAQ/Sick Building Syndrome Problems, in General, and Some Solutions

An IAQ problem can be of natural origin (i.e., radon), or, it can be due to various VOC emissions from different indoor sources, mostly associated with an inadequate supply of fresh air. Deterioration of IAQ eventually leads to the sick building syndrome. Hence, availability of fresh air is very important in combatting this problem. In principle, a house should have a complete air change regularly. A typical old house is expected to have more frequent air changes due to possible leaks already existing (natural stack effect). In any case, the need for fresh air can be decreased if more electric heat and heat pumps are used in place of gas furnaces and gas water heaters in the house. If heated by gas, an airtight house can have carbon dioxide levels two to six times higher than outdoors.

Installing a heat recovery ventilator (HRV) is the most efficient method to bring in fresh outdoor air all the year-round. The small circulation fans use around 100 watts of electricity and it is much more efficient and controlled than just opening windows or relying on the wind to blow. HRV incorporate one of several designs of heat exchanger cores. In the winter, cold incoming fresh outdoor air picks up heat from warm outgoing stale indoor air. The stale indoor and fresh outdoor air paths are sealed from one another, so no pollutants are transferred. HRV save from 65% to 80% (efficiency) of the energy from the outgoing stale air. In the summer, the air flows through the same heat exchanger, by the heat flowing in the opposite direction. The cool outgoing stale air pre-cools the incoming fresh air. Small window units are very easy to install and these lightweight units are mounted in windows just like an air conditioner and can be moved from room to room. Larger through-the-wall models are also available and easy to install. With HRV, the fresh air gradually circulates throughout the entire house. They operate rather quietly as many internal components are attached by rubber mounts.

Heating and air-conditioning systems keep buildings warm in winter and cool in summer, however, they do not help to improve air quality in the house. A total heating-ventilation and air conditioning system (HVAC), which should include a furnace, an air filter, humidifier, make-up-air unit and air conditioner, is very effective in improving IAQ. If the HVAC system used is not total, it can cause the circulation of harmful, contaminated air throughout the home, and hence can activate sick building syndrome. If the HVAC system does not effectively distribute air to people in the building for some reason, then inadequate ventilation is found and, hence poor IAQ results [4]. In fact, older office-HVAC systems were designed for 'one person per 15-20 m^2, and a PC on every third or fourth desk' and in a modern office with higher occupant densities, it is 'for more people in much less space'. Today's standards require 0.57 m^3 per minute (approximately 10,000 litres) of outside air per person.

Plants are found to improve and cleanse the indoor air from a number of harmful VOC pollutants such as formaldehyde, benzene and trichloroethylene, as shown in a NASA study. Golden Pothos (*Pothos aureus*), *Philodendron*, Corn Plants (*Dracaena Fragrans*), and Bamboo Palms (*Chamaedorea seifrizii*) are found to be effective in cleansing the air from formaldehyde. Peace Lily (*Spathiphyllum wallisis*) and Janet Craig (*Dracaena deremensis*) are good for removing small quantities of benzene. Trichloroethylene can very effectively be removed by *Dracaena marginata*, Warneckei (*Dracaena deremensis*) and *Spathiphyllum*. It is recommended by the Plants for Clean Air Council that there is one potted plant for each 10 m^2 of floor space for better IAQ [5].

7.2.3 Four Elements of Sick Building Syndrome

There are four elements of sick building syndrome in general:

a) ***Chemical contaminants from indoor sources*** are probably the predominant direct source of indoors air pollution. There are a number of synthetics, adhesives, carpeting, flooring, upholstery, manufactured wood products, various construction materials, in addition to copy machines, pesticides, and cleaning agents that all emit VOC. If coupled with poor ventilation, they can create poor air quality, which is believed by adherents of sick building syndrome to either create health problems or exacerbate pre-existing ones.

b) ***Chemical contaminants from outdoor sources*** are more indirect than indoor contaminants. Pollutants such as motor vehicle exhausts can be conveyed indoors through air intake vents, doors, and windows.

c) *Inadequate ventilation* can occur when heating, ventilating, and air conditioning (HVAC) systems do not effectively distribute air to people in the building, as discussed previously.

d) *Biological contaminants* are bacteria, moulds, and viruses. These contaminants may breed in stagnant water that accumulates in duct work, humidifiers, and the like, or where water has collected on ceiling tiles, carpeting, or insulation. Insects or bird droppings, too, can be a source of biological contaminant. Physical symptoms related to biological contamination include cough, chest tightness, fever, chills, muscle aches, and allergic responses such as mucous membrane irritation and upper respiratory congestion.

These four elements may act in combination, and may supplement other complaints such as inadequate temperature, humidity, or lighting.

There are studies to model the sick building syndrome in residential interiors depicting the relationship between common health problems and factors leading to sick building syndrome [6].

There are studies to suggest safer construction materials with the lowest possible risk on health for the indoor environment, mainly to decrease the effect of second factor (chemical contaminants from indoor sources) listed above [7, 8]. Most of these are natural', or in other words, they are 'green'. However, one should also consider the fact that, all natural things are not safe. A number of natural materials can contain VOC and hence can pose hazards to health as well. Radon is an example, it is a natural material, which is radioactive and exists almost everywhere in the house. In addition, allergic reactions to the odours from cedar furniture, a natural material, are very common. The risk is always low if a certain agent remains in the building product that does not affect occupants through respiration and physical contact.

Less toxic (natural or synthetic) alternative materials should be used whenever possible, and if this is not possible, than products that off-gas a little should always be used in preference to those that off-gas a lot.

7.3 Volatile Organic Compounds (VOC)

7.3.1 Possible Sources of VOC

A volatile organic compound is a material that at normal (ambient) temperatures or under the influence of heat, is capable of being vaporised or becoming a gas (i.e., solvents used in paints). Some materials indoors may continue to generate VOC over many years, the concentration of which varies with variation in temperature, airflow and volume of the house.

Possible sources of VOC indoors will be outlined later in Section 5.

7.3.2 Permissable Limits for VOC Indoors

As far as the permissable limits of VOC concentrations indoors is concerned, there is no universal regulation established yet, (i.e., in the US, there is no such federal regulation existing; and several regulatory agencies such as the EPA and the Occupational Safety and Health Administration (OSHA) worked on developing several standards. However, none of these are easy to apply because the correlation between methods and indoor VOC concentrations is not straighforward. In addition to the fact that detection of specific low VOC at low concentrations 'out-gassing' may not indicate whether there will be any long-term negative health effect or not.

However, it is still be essential to have VOC emissions information for any material that is to be used in any construction to make proper decisions on which materials best meet the requirements while fulfilling structural and aesthetic needs.

In a compherensive study for IAQ and sick building syndrome of certain office buildings selected in the US [9], it is reported that total VOC (TVOC) can range from 73-235 µg/ m^3 where the most prevalent compounds were xylene, toluene, 2-propanol, limonene, and heptane. Formaldehyde concentrations (geometric mean) are found between 1.7 to 13.3 µg/m^3 and mean aldehyde levels from <3.0 to 7.5 µg/m^3. A number of upper respiratory symptoms (dry eyes, runny noses), symptoms of central nervous system (headache, irritability) as well as musculoskeletal symptoms (pain and stiffness in neck) were found to be high within the workers in the office studied. While in another study in Japan (to assess the impact of office equipment on the IAQ), it was found that the emission of ozone and organic volatiles (mainly formaldehyde followed by lesser amounts of aldehydes) emitted are in significant quantities [10]. In one application, in the Washington State 'East Campus Plus project' in the US, office furniture systems were required to emit no more than 0.05 ppm formaldehyde and 0.50 ppm total VOC to be considered for their installation. While in an another study in Finland [11] estimation of the impact of office equipment on IAQ was questioned and the emission of ozone and various organic volatiles from photocopiers and laser printers were determined. The laser printers equipped with traditional (corona rod) technology were found to emit significant amounts of ozone and formaldehyde, with lesser amounts of other volatile aldehydes, and it is suggested that these are not to be placed beside or immediately at the working site of office personnel.

To give some more depth on this subject, some basic concepts of toxicology and toxics are summarised in the Chapter 2.

7.4 Risk Management and Some Notes on Toxic Compounds that can be Found in Indoor Spaces

7.4.1 Risk Management

Since any material must be 'assessed' in the context of the system, there is no truly benign materials and nothing is risk free and 'risk can be managed', i.e., a toxic material can provide significant benefits and may pose little risk 'when used properly'. For example use of damp-proofing on the exterior of a preserved wood foundation that has an inherently toxic chemical provides moisture control and substantial benefits, but does not pose a high risk to the occupants [7].

Overall risk assessment rests on three factors:

a) Exposure assessment,

b) Toxicity assessment, and

c) Dose-response assessment.

Exposure assessment is a necessary component in understanding the hazard involved by exposure to naturally (i.e., radon) or non-naturally existing toxicants (i.e., chemicals emitting from construction materials) [12]. However, the other two (toxicity and dose-response assessments) are the next two important factors to know.

Table 7.1 shows some toxic compounds that can commonly exist indoors, mainly due to natural and synthetic construction materials [13]. In addition, radon is added to the list, although it is not directly related to construction materials, but a very important item indoors, but yet not well known.

Certainly one should consider the fact that this is a general list and each indoors is unique and a specific indoors may have always have different toxics.

7.5 Some Notes on Toxic Materials that can be Found Indoors

Tables **7.1** and **7.2** present some toxic compounds that can be found in indoor spaces originating mainly from construction materials (**Table 7.1**) as well as their possible effects on humans (**Table 7.2**).

Table 7.1 List of some toxics that can be found indoor spaces	
Toxins	Indoor Spaces
Radon	A naturally occurring radioactive gas, leaking from basements, crawl spaces and water supplies.
Formaldehyde	Mainly from particle-board (and furnishings) and other aldehydes.
Other VOC such as: Aliphatic hydrocarbons – e.g., mainly hexane	Emitting from carpets, flooring materials, paints, furniture.
Halogenated aliphatic hydrocarbons – e.g., mainly chloroform	Found in natural and synthetic resins, paints and lacquers
Aliphatic alcohols – e.g., methanol, ethanol and 1-butanol	Found in paints, adhesives and pesticides
Glycols and glycol ethers	
Aromatic hydrocarbons – e.g., mainly benzene, toluene and xylene	
Various endocrine disrupters (ECD): (certain additives and plasticisers)	

Table 7.2 Effects of some of the VOC toxics found indoors	
VOC Toxic Compounds	Effects
Benzene	Mainly cause dizziness, headache, vomiting, drowsiness and unconciousness at low doses. Chronic exposure - if contact with the eyes: neuritis, atrophy, visual impairment, oedema, and cataracts, can cause CNS depression, bone marrow depression, leukaemogen headaches, anoexia, nervousness, weariness, anaemia, pallor, reduced clotting, marrow damage and finally leukaemia. Deliberate inhalation of benzene vapours (glue sniffing) can kill directly.
Chloroform	Chronic exposure - can cause liver, kidney, nervous system disorders and heart damage. It is a carcinogen and gives rise to alteration of genetic material (please see Section 7.5.1 for ECD), fertility problems, foeto-toxicity and developmental problems such as craniofacial, musculoskeletal and gastrointestinal.
Radon	Radon is a chemically inert, and a highly unstable radioactive gas from the radioactive decay of uranium-238 of natural origin, which is colourless, odourless and tasteless, which tends to accumulate in buildings and can pose a serious risk to the health, causing lung cancer, if its concentration is high [14-16]. For more detailed information about radon, please see the Appendix (A-7).

149

7.5.1 Endocrine Disrupters (ECD) and Some Suspected ECD Agents Indoors

As outlined in Chapter 2, ECD are chemicals that can cause 'hormonal related diseases' and 'dysfunctions' that can be effective even at very low levels (at parts per trillion, levels at which most chemicals have never been tested). ECD became a significant focus of environmental science and medicine in recent years because of its critical importance in heath. A wide range of chemicals, both natural compounds (phytoestrogens) and some synthetic chemicals (including polychlorinated biphenyls (PCB), dioxins, certain preservatives and metal ions, even certain woods) are all suspected of being capable of endocrine disruption in humans.

There are four main groups of chemicals that are labelled in general as 'suspected ECD agents':

- Certain plastics and rubber additives,

- Certain PCB,

- Chlorinated dioxins and dibenzofurans, and,

- Certain metal ions and metal compounds.

7.5.1.1 Certain Plastics and Rubber Additives (Mainly Plasticisers) as Suspected ECD Agents in Houses

Some plastics additives, such as phthalates, bisphenol-A (BPA), and nonylphenols, present as plasticisers, can evaporate indoors and can be inhaled i.e., oestrogenic butyl benzyl phthalate is found in most vinyl floor tiles, adhesives, and synthetic leathers, and bisphenol-A is a breakdown product and plasticiser of polycarbonate (PC) mainly used as glazing material. In addition to plasticisers, there are also other additives used in plastics and rubbers i.e., stabilisers used in PVC window profiles and pipes are mostly lead-based, or they can be either barium/cadmium/zinc compounds. All of these can pose serious health hazards if they are above certain concentrations in the air, through evaporation from the system. Increase of temperature indoors speeds up this evaporation process.

a) Some Common Plasticisers in Plastics and Rubbers used Indoors which Act as ECD Agents

PVC dominates most of the literature on external plasticisers, mainly phthalate esters of C_8 to C_{10}, used to flexibilise PVC indoors flooring products to make them easy to roll, store and install. Some external plasticisers have high and low volatilities, but they are mostly toxic and ECD agents [17], and there is the probability that they will be emitted and accumulate indoors, over time. In addition, they can also exist in the household dust whenever they are abraded in use (i.e., from plasticised vinyl floor surfaces), and even during washing of vinyl floors, phthalates can get into the environment. Certain analytical techniques are available to detect traces of most plasticisers indoors (at the parts per billion (ppb) level) [18].

The harmful effect of plasticisers indoors was first realised in the 1980s, when it was seen that vapour emitting from dibutyl phthalate (DBP), and plasticised PVC (PVC-P) glazing seals were damaging certain greenhouse crops. PVC-P used indoors, on average, contains 55 phr (parts per weight per hundred parts of PVC) plasticisers, which can be di-*n*-octyl phthalate (used in the manufacture of flooring and carpet tiles), diethyl hexaphenyl (DEHP) (in PVC floor coverings, recently replaced by another phthalate, di-isononyl phthalate (DINP)), di-isodecyl phthalate (DIDP) (mainly in wire and cable production and carpet backing applications), DINP, and butyl benzyl phthalate, (BBP) (mainly in vinyl tile production), and di-*n*-hexyl phthalate (DHP) (in flooring applications). Five of the phthalates (dibutyl phthalate (DBP), DEHP, DINP, DIDP and BBP) are currently undergoing EU risk assessment.

Polyvinyl acetate (PVAc) (as adhesive), as well as cellulose acetate (CA) compounds and sheets, cellulose nitrate pigment binders and polyvinyl butyral (PVB) sheets (used mainly for safety glass interlayers) are the other main users of a range of different plasticisers.

There are also several external plasticisers with almost no proven toxicity, such as, tri-(2-ethylhexyl)trimellitate (TEHTM), di-2-ethylhexyl adipate (DEHA), and acetyl triburyl citrate, (ATBC), which are economically unfeasible for industrial applications, i.e., TEHTM is three times (and for DEHA, four times) as expensive as DEHP. In any case, the use of adipate, mellitate and azoalate type external plasticisers are expected to grow in use at the expense of different phthalate types [19]. Butene based alcohols are also used in the manufacture of flexible PVC [20], whereas polycaprolactone is used as a permanent and safer plasticiser for PVC [21].

For rubbers, process oils, which are simply some hydrocarbons, function to plasticise the system (polybutenes are plasticisers in polybutadiene rubber based membranes used for roofing systems).

As mentioned previously, the EU put forward the year 2002 as the key milestone to complete phthalates risk assessment, although this wasn't actually achieved. Currently, it is known that there is a high interest in plasticisers and their effects on health worldwide, and that in the EU, about one million Euro a year is being spent on such research in industry.

For additional information about plasticisers and their possible health effects, please see Chapter 3.

b) *Some Common Stabilisers, Flame Retarders, Pesticides, Antimicrobials in Plastics and Rubbers used Indoors as Effective ECD Agents*

The common stabilisers for PVC are compounds of lead (lead sulfate and lead stearate that are economical in use, but are highly toxic), tin (mono- and dibutyltin as well as thioglycolate, with excellent thermal stability and low toxicities), cadmium (toxic, causes kidney damage and anaemia - their phasing out has been underway) and complex salt systems of barium/zinc and calcium/zinc. Lead systems, although considerable toxicities involved, are still expected to remain the dominating stabiliser type until any legislation dictates otherwise. Recently pyrimidine-dione type stabilisers with no heavy metals have been introduced. Hindered amine light stabilisers (HALS) are used as scavengers, in addition to some organo-nickel compounds as quenchers, for UV stabilisation.

Pesticides and antimicrobials (biocides) are used in a number of plastics construction materials, mainly to provide resistance to the growth of microorganisms. One common example is their use in a number of PVC and PU grades. PU is commonly used for roofing membranes in construction [22], and several plasticisers, lubricants, thickening agents and fillers that are used with it can support microbial growth effectively.

Use of these materials in high humidity atmospheres can activate microbial attack. Some commonly used plasticisers (DOP, DIOP, DBP, tricrescyl and TPP triphenyl phosphate (TCR) are the most resistant to microbial attack. The major antimicrobial agents used in PVC are 10-10′-oxybisphenoxarsine (OBPA), *n*-(trichloromethylthio)phthalimide and 2-*N*-octyl-4-isothiazolin-3-one (OITO).

Within the flame retarders, there are the halogens (chlorine and bromine being the most effective) and phosphorus containing compounds, and there is a synergy found between antimony, zinc and other metal salts. Within common flame retarders, they are mainly hydrates, such as antimony trioxide and aluminium or magnesium hydroxide, alumina trihydrate, zinc borate, phosphate esters and chloro-paraffins. Most of these were developed after the ban on halogen containing flame-retardants. Most of these are volatile to a certain extents, hence can be dangerous for indoor use. The ban on halogens is because of their toxic nature, and especially because of their gas emission when the system is heated, however, the pressure to ban in Europe has abated a lot nowadays. 'zero halogen' flame-retardants are mainly used for cable applications.

Chlorinated paraffins (mainly CPVC) are widely used in PVC to have greater resistance to ignition and combustion than general purpose plasticisers. However, the effects of chloroparaffins on health is still a controversial issue and their use as flame retarders in PVC applications for cables, wall coverings and flooring are declining [23]. For more information and health effects of stabilisers, flame-retarders, please see Chapter 3.

7.5.1.2 Indoors Atmospheres and Suspected ECD Agents

A Swedish study showed that, there are still high levels of PCB around some old buildings (where certain sealants were being used 20-40 years ago). These sealants were based on polysulfide polymers used for filling external joints in buildings from the 1950s and until the late 1970s, and they usually contain up to 20% PCB. In fact, a very high PCB level - about 100 times the typical ambient levels - in Stockholm was measured on a balcony on a hot summer day. Although these sealants are normally not used inside buildings, the study in one exceptional case found high levels of PCB in the stairway of a building too. The report recommends checks on PCB levels in all structures built between 1956-1972. PCB sealants have also attracted attention in the US, UK and Germany, but no research or monitoring is in effect. PCB in many polysulfide sealants have been replaced now by chlorinated paraffins, which also have certain restrictions raised by the Oslo and Paris commission [24, 25].

7.5.1.3 Indoors Atmospheres in Houses and Certain Metals and Metal Compounds as Suspected ECD Agents

It is known that, when certain metal compounds are present in air at concentrations above the levels of homeostatic regulation, they can act as a health hazard. The most common metals and metal compounds that can be found in the indoors atmosphere, are mainly those of antimony, lead, (methyl) mercury, and cadmium, which can exist in plastics as additives (mostly found and used as stabilisers). These are believed to disrupt the endocrine system by causing problems in steroid production. The fate of these metal and metal ions has been more extensively studied for the lead and lead-based compounds, however, other metals have been studied much less than others.

Lead (Pb) compounds, the deadly cumulative poison indoors, can leach from old water pipes (in the form of soluble lead), and from badly glazed pottery, and even from lead crystal decanters. In a study, it was shown that the rate of lead extraction from a 100 mm diameter PVC waste water pipe system was 0.7 µg [26] and that the sewer system can contribute 0.5 µg/l/lead to the wastewater [27]. On the other hand, in the CSIRO report it is concluded that, 'under normal use conditions in the potable water industry, the level of lead extracted from properly commissioned PVC pipe has been found to be below the levels of detection' [27].

Some old paintings in buildings may also contain lead.

Lead or lead compounds are absorbed into the body by inhalation. If the amount absorbed is small, the body can get rid of some of it through urination, and some may still stay in the body, stored mainly in the bones - this can stay there without any poisoning effect, until a certain dose is reached by accumulation over time [28].

The EU initiated studies to control and stop completely the use of lead stabilisers in plastics by the year 2015.

The source of antimony compounds indoors is mainly PVC (used as a flame-retardant additive). During the 1990s, antimony containing PVC sheets were thought to cause cot deaths in babies (it was claimed that antimony was converted to the volatile toxic gas stibine easily by a fungus existing in the mattress). However, this claim has not been proved so far and although the analysis of tissues from cot death victims had antimony levels somewhat higher than allowed (13 ppm), similar results were also observed for healthy babies. Moreover, in the house dust of some old houses the level of antimony can already exceed 1800 ppm, the source of which is not exactly known.

Another deadly poison, arsenic [29] can emanate mainly from paintings (the bright yellow pigment, called 'royal yellow' is in fact 'orpiment' or arsenic trisulfide, which was favoured by Dutch painters in the past-this slowly oxidises to arsenic trioxide, which is extremely toxic, and causes fading of the colour). Arsenic is also used in treated lumber, and can exist indoors in lumber.

Cadmium compounds are a cumulative type of poison, and they are used as the common bright pigment (cadmium yellow) in the form of cadmium sulfide in paints, and also in rubber and plastics as an additive. Until recently, cadmium red was widely used for containers, toys and household wares (which have been completely phased out now). From March 2001, sales and use of any cadmium-containing stabiliser was banned in EU.

As phosphorus-containing compounds are mostly used as flame-retardants (specifically in the synthetic fibre industries, such as polyester production), it is not surprising to find its compounds in the indoor environment.

Tin compounds indoors can be most dangerous if they are ingested (organotin compounds like trimethyl tin (TMT) or triethyl tin (TET). Tin (with one or two organic groups) are effective additives used to decrease heat sensitivity of plastics (as organotin stabilisers for example in PVC while tin (with three groups attached, such as tributyl tin) are widely used as wood preservatives and in anti-fouling paints, as well as to prevent unwanted growth of moulds on stone structures. Organotin compounds (with four organic groups attached) are used as catalysts in the production of polyolefins (such as polyethylene (PE)).

Zinc is non-toxic in general, however, a surplus of it can be stored in the bones and spleen, and the most significant toxic effect of zinc is fume fever, that can result from acute inhalation of zinc oxide fumes. Zinc oxide is used in the rubber industry (it is a catalyst during manufacture and also used as a heat dispenser in the final product), as well as in pigments for plastics and in wallpaper. It also functions as a UV stabiliser for plastics and rubbers. In addition to these, it is possible to find mercury vapour indoors, emitted mainly from the biocides used in paints.

More detailed information on these metal ions regarding their effects on health are provided in Chapter 10.

7.5.2 Effect of Some Plastics, Rubbers and Wood-Related Materials on the Indoors Atmosphere in Houses

Some of the ingredients, mainly additives existing in various construction materials can slowly evaporate (as outlined in Section 7.5.3), and/or breakdown, meanwhile releasing different chemicals. In construction materials, plastics and wood are our main concern.

7.5.2.1 Some Thermoplastic Construction Materials (PVC, Polymethylmethacrylate (PMMA), Polyolefins and PC) used Indoors

The main use of PVC is mainly for piping and pipe fittings (pressure piping for potable water as well as for gas and normal piping for drains and sewerage), cable and wiring covers, electrical switches, conduit, roofing/building membranes, insulation, flooring, wallcovering, trim, carpet fibre and backing, miniblinds and shades, window frames, doors and siding, partitions, and foams. PVC can have a long life of between 15-100 years.

Most of the additives used in PVC can pose several health problems [30, 31], as outlined previously, and this has been a subject of a very intense debate for many years, beginning with the 'danger of release of toxic stabilisers, phthalate plasticisers and other toxic additives' to 'the danger of formation of dioxins and hydrogen chloride gas in the case of fire'. In fact, when it is burned (i.e., in a house fire), PVC-P is claimed to produce 'dioxins', which are known to be a deadly poison and a strong carcinogen, plus the corrosive and toxic hydrogen chloride gas. Routine PVC thermal welding revealed vinyl chloride monomer (VCM), benzene, formaldehyde and acetaldehyde [32]. Overheating PVC can cause acute upper and lower respiratory irritation (due to toxic hydrogen chloride and carbon monoxide emissions [33]).

For many years, there has been a never-ending debate between different parties about PVC and on its effect on health and on environment. Some of these theories are correct but most of them are speculative and opportunistic [34].

PVC is produced from its monomer VCM. VCM, being a rather small molecule can readily vapourise and become fatal. It is highly carcinogenic and can cause liver and brain cancer [35, 36]. VCM can stay in the polymer in trace amounts after its production, and even after the processing of PVC into its final shaped products. The 1997 European Pharmacopoeia requires a maximum of 1 ppm of VCM residual in virgin PVC. As a final note, it is worth mentioning that there is a EU voluntary commitment study on the PVC industry initiated in 2001 for the following 10 years (called Vinyl 2010), including mid-term revisions of targets in 2005 and definition of new objectives in 2010 [23, 34].

Polycarbonates are mainly used as a glazing material in construction (transparent roofing, impact-resistant glazing, sheet and for structural parts in building and construction). Greenhouses and the dome of the Sydney Olympic stadium all have PC sheet glazing. PC sheets are virtually unbreakable (bullet resistant windows, protective PC glazing panels). PC resins and BPA are known to be safe and they pose no health risk to humans.

The building block for PC is BPA. This monomer exhibits toxic effects only at very high exposures and realistically, such high exposures are not possible under normal conditions indoors. BPA is not a carcinogen or a reproductive or developmental toxin.

Polyolefin is a generic name for PE and polypropylene (PP). The burning of these plastics can generate several volatiles, including formaldehyde and acetaldehyde, both of which are suspected carcinogens.

PE has a very low compatibility with plasticisers and in fact it does not need plasticisers, however, they may contain other additives (i.e., UV and heat stabilisers). Chloroparaffins or brominated flame-retardants are common used in polyolefins.

Low-density PE, medium-density and high-density (HDPE) are mainly used for piping (mainly for pressure pipe production) and floor coverings in the construction industry. HDPE is commonly used in perimeter drain pipe around foundations, but rarely inside houses.

Corrugated HDPE pipes are used in mortar, walls and concrete.

PE foams (expanded PE or EPS) are used in various applications for seals and insulations (on exterior walls, interior or between the walls, in flooring and hot water pipe insulations) in building and construction.

Polyolefin floor coverings (PP and PE), power cables with PE coverings and HDPE pipes and wall covering materials, halogen free linear low-density PE and thermoset crosslinked polyethylene are all suggested as alternatives to PVC by Greenpeace.

PMMA (or Plexiglas, also known as acrylates or acrylics) is made from the monomer methyl methacrylate (MMA), and are known for their excellent optical clarity, colour stability, and good weatherability characteristics. They are mostly used for glazing, lighting, curtain-wall panels as a sealant, and for decorative features. PMMA can contain some of its monomer, MMA, which can also be evolved from thermal degradation of PMMA. MMA is toxic.

Polystyrene (PS) can contain some of its monomer, styrene, which is toxic (Group C - possible human carcinogen) and is the only source of toxicity of PS. PS is mainly used in construction in the form of high performance EPS foam, as insulation for floors, walls and roofs. PS foam, (known as Styropor, BASF) is mainly used for thermal insulation of buildings. For EPS, styrene monomer is used which is known to be toxic to the reproductive system, and hence residual monomer is a problem with the use of EPS as well. The burning of bulk or foam PS releases styrene as the hazardous chemical.

7.5.2.2 Some Thermoset Construction Materials (Polyesters, Epoxides, PU and Phenolics) used Indoors

Epoxy resins are the most commonly used thermosets, used for durable and inert coatings, in laminates and various composite applications, as well as an adhesive. Epoxies have low vapour pressures, so the potential for respiratory exposure is very low, and hence they do not pose health hazard indoors. The potential for dermal (contact) exposure is, however, much greater than respiratory exposure, which should not be a problem at all for a finished construction indoors. It is a reaction product of epichlorohydrin and BPA, which are not carcinogenic or a ECD agent.

Its curing agents also have low vapour pressures and so do not present any airborne hazard.

PU in the form of foams, can be rigid (used mainly for load bearing applications and for insulation in construction) or flexible (used for upholstery and as carpet backing). Usually, rigid foams with densities greater than 320 kg/m^3 are called 'structural foams' [37], and are used as substitutes for wood, metal or unfoamed plastics. PU are formed by reacting a polyol with an isocyanate compound. Polyols are relatively safer, but isocyanates are highly toxic in general, which can represent a significant respiratory hazard indoors, if some of the ingredients are assumed to remain in the foam system after its processing. Exposure to the vapour may cause severe irritation of the eyes, to the respiratory tract, severe enough to produce bronchitis and respiratory *sensitisation* (an allergic, asthmatic-type reaction) at high concentrations. Catalysts for PU foams (tertiary amines and organometallic compounds, particularly organotin compounds such as dibutyltin esters) have high vapour pressures, and are considered to be safer. However, tertiary amines can still act on the respiratory system as a poison.

Foamed cellular plastics provide a number of advantages, however, they can also bring two important problems as well. These are, the problems of 'the residual chemicals that may be left in the system' and the problem of the 'flammability and smoke evolution on combustion'. In addition, there can certainly be residues of unreacted monomer(s), blowing agent(s), polyols, catalyst(s), fire retardants left if the system is not prepared carefully, all of these can be left or trapped in the system (inside the internal cells). These can be released slowly with time by themselves, or as the cells are broken down in use, with the possibility of them causing some additional health problems.

Foams, especially PU foams, present considerable fire hazards. To suppress flammability, fire retarders are added to the system. In addition to the foam itself, some flame-retardants i.e., halogen containing ones, contribute a considerable amount of smoke, so that smoke evolution can become a far greater danger to people than the fire itself. Flame retarded rigid PU foams, or PU in bulk, when ignited at high temperatures, can still give off highly poisonous cyanic acid (12 ppm) and small amounts of carbon monoxide gases indoors, even dioxin [38].

Phenolic and amino resins have the well-known hazards, due to both phenol and formaldehyde. In addition to traces of free formaldehyde, they may also have free phenol, which is a known toxic.

The urea – formaldehyde (UF) and melamine–formaldehyde systems represent similar hazards. Free formaldehyde, which can be present in trace amounts, may be liberated to the air when resins are processed or even slowly afterwards, which can irritate the mucous membranes. Formaldehyde is a metabolite occurring normally in the human body and is converted to formic acid by enzymic oxidation. Formaldehyde in the cured resin is believed to be due to the unreacted free formaldehyde left (there are also claims that it may be due to demethylolation reaction and/or cleavage of methylene-ether bridges as well). A model specification to set out the health hazards is presented for polymer mortar surfacings (out of epoxy, polyester and PU thermosets), intended for their use as indoor floor tappings [39].

7.5.2.3 Rubbers and Indoors

Rubbers (or elastomers) are hydrocarbons (or silicones) and their application is mainly in floorings, sealings and roofing. Rubber flooring with chlorine-based ingredients are not recommended because of the health hazards involved (danger of hydrogen chloride gas evolution). For other rubber types, such as ethylene propylene diene terpolymer (EPDM) rubber, there is no such danger involved (EPDM is recommended by the Danish Environmental Protection Agency as an alternative to PVC).

The rubber industry uses hydrocarbon additives, specifically called process oils (mineral oils, functioning more or less as plasticisers, below 20 phr) or extenders (used for economy). Polycyclic aromatic hydrocarbons, containing process oils and with high degrees of vaporisations, are classified as carcinogens. In EPDM formulations, (i.e., for roofing membranes), liquid polybutene, a safe compound, is used as process oils. The term plasticiser for rubber, referres to the special synthetic liquids used with the polar rubbers (i.e., triethylene glycol di-2-ethylhexanote and similar esters, which are the most favourable low temperature plasticisers for polar synthetic rubbers).

As for rubber gaskets used indoors, those made of EPDM rubber have been tested indoors for their emissions in Sweden in an experimental wooden house at the Swedish National Testing Institute [40].

Most rubbers burn easily, producing considerably high heat and acrid smoke, which usually contain harmful constituents, including halogens in a lot of cases. Self-extinguishing properties for rubbers can be obtained by appropriate compounding.

7.5.2.4 Wood, Wood Products, Wood Laminates and Indoor Atmosphere

Wood by itself, as well as its composite products, such as particleboard, plywood, and medium density fibreboard (MDF) - are widely used in indoor products in the construction sector (as structural panels, sub-flooring, ceilings, door cores, as well as in cabinets, panelling, furniture, and so forth). Particleboard and MDF are generic terms for panels composed of cellulosic materials (primarily composed of lignocellulosic fibres combined with a proper synthetic resin or bonding system bonded together under heat and pressure, respectively).

After the introduction of processing facilities for wood-plastic composites (WPC) by extrusion and later by injection moulding techniques, WPC production and its use increased almost exponentially.

Most wood products are simple combinations of wood and certain water-based adhesives (mostly of UF or phenol-formaldehyde (PF) plus catalysts, filler and extenders - for plywood products). In some special applications, products can also be bonded with methylene diisocyanate. These chemical resins used as adhesives were blamed for a problem of emission of various toxics. Research on VOC emissions from wood products has so far focused mainly on formaldehyde, although formaldehyde is given in List 3 of EPA as 'a chemical of unknown toxicity' and phenol in the List 1 as 'chemical of toxicological concern'. The particleboard industry, primarily through new resin technology and better process control techniques, has reduced formaldehyde emissions to very low levels. Permissible limits of formaldehyde emission show differences from country to country (American National Standards Institute (ANSI), restricts this emission from particleboard flooring to 0.2 ppm and for others to 0.3 ppm. OSHA has adopted a permissible exposure level of 0.75 ppm and an action level of 0.5 ppm, some states established a standard of 0.4 ppm. In their codes for residences, others, i.e., California, established a lower level of 0.05 ppm, while in Germany, it is 6.5 mg/100 g and 7.0 mg/100 g of dry particle board and dry board for MDF, respectively, which corresponds to 0.1 ppm also recommended by WHO). Melamine laminates commonly used for kitchen and bathroom cabinets, door facings and countertops may also have residual formaldehyde volatiles, from either the laminate itself, or the substrate material. In fact, formaldehyde is normally present in the air around us, however at low levels (<0.03 ppm). Typical levels of formaldehyde in the smoking section of a cafeteria, is approximately 0.16 ppm.

Tests with composite wood products have also yielded volatiles of different preservatives (in addition to formaldehyde), such as: different volatile solvents or free monomers and plasticisers from adhesives that were used during its production (commonly phenol or formaldehyde-based adhesives are used in the manufacture of compressed fibre, composite board and plywood materials) or from coatings applied to them (some products that are sealed with a polymeric film or coating which can trap residual volatiles and allow a slow gradual release of VOC over a period of time). The binding agents used in particleboard and plywood can contain volatile phenols and traces of residual solvents. In the laminate application of wood (particularly on cabinets and cupboards that are made from porous composite wood materials such as particle board), it is always possible to have trapped solvents (mainly aromatic hydrocarbons such as toluene and xylene, and ketones such as acetone and methylethyl ketone (MEK) in the adhesives. The adhesive solvents that are absorbed at the wood surface can diffuse into the air, in standing over long periods of time, although in small amounts. Xylene and toluene are given in the List 2 of EPA as 'potentially toxic'. Formaldehyde is a strong irritant to mucous membranes. Adverse health effects that are associated with increased VOC concentrations of formaldehyde can begin with eye and respiratory irritation (including allergy and asthma), irritability, inability to concentrate and sleepiness, and can end up with more serious health disorders. EPA tested formaldehyde levels in a newly constructed test home that contained various UF-bonded building materials have found values below that of the expected level (0.06 ppm), and it is not clear whether adsorption of formaldehyde, most notably by painted gypsum wallboard, has contributed to this unexpected result or not [41].

In addition to the improvements involved in resin technology and better process control techniques, there are different methods applicable to producing wood products with lower (minute amounts of) formaldehyde emissions as well. To recap: firstly, resin formulation can be changed or ageing can be applied to it - emissions decay considerably with time so that within a year or so, their level can be as low as the background ambient levels.

Recent studies revealed that, in addition to aldehydes in such systems, there may also exist terpenes, ketones, acetone and hexanal [42-44]. This finding has initiated a growing interest for 'other possible non-formaldehyde based VOC emissions' from wood and wood products.

Wood furniture coatings usually contain urethane/isocyanates, in addition to UF, volatile plasticisers, residual solvents from application and free monomers from incomplete polymerisation of the coating. Nitrocellulose lacquer, acrylic, cellulose acetate butyrate and PU (with plasticisers like epoxy, DBP, BBP and isopropyl myristate) are the coatings commonly used on clear finished wood furniture. In these, the amounts and the types of VOC mainly depend on the type of curing. Heat cured and high solids coatings usually retain the lowest amounts of organic solvents. In one study, emission characteristics of the release of VOC, especially detection of photoinitiator fragments, from UV cured furniture coatings was studied, and it was found that they contribute significantly (20-60%) to the total emissions [45]. One main compound found in this study was benzaldehyde, generated by many applied photoinitiators via cleavage. While in another similar study [46], environmental issues indoors related mainly to wood products and furnishings were surveyed.

There are also applications with wood composites prepared with thermoplastics (PE, PVC, PP and polyethylene terephthalate), and 'plastic lumber concept' is evolved as a result of this concept [47], which is developed mainly for their use outdoors, but use in indoors are also possible [48]. In general, certain coupling agents are applied in such systems to enhance their performance.

Since the 1940s, lumber producers and manufacturers have been using a chemical compound mixture that contains inorganic arsenic, copper, and chromium, called chromated copper arsenate (CCA) as a wood preservative. CCA is usually injected into wood by a high pressure process to saturate wood products with the chemicals, to produce 'pressure-treated lumber' (between 75 and 90% of the arsenic used in the US is estimated to be used for wood preservation).

7.5.3 Some Construction Applications and Related Possible Health Hazards Indoors

In following sections, some construction applications of polymers will be summarised. Each construction material should be considered first for the possible inherent hazard(s) involved, which were presented before, if they are not specifically mentioned next.

7.5.3.1 Asbestos

Before the mid-1980s asbestos was one of the main ingredients of the 'resilient flooring' and 'acoustic ceiling tiles' industry. Asbestos is a mineral and it is hazardous to health when it becomes friable or free floating and airborne indoors, as in a dust. Asbestos has been banned for a long time and it is no longer used in construction. But in a building older than the mid-1980s, it can still exist as an insulating coating on steelwork or concrete, as lagging on pipes and boilers, as insulation boards in walls, on doors and ceilings, as asbestos cement for roof and wall coverings, pipes and tanks, and in some decorative plasters. The danger comes from drilling, cutting, sanding or disturbing materials made from asbestos and then by breathing the dust. The EPA has determined that encapsulated or nonfriable asbestos containing products are not subject to extensive regulatory requirements as long as they remain in that state (provided that they are not sanded, sawed or reduced to a powder). There are still a number of people that suffer each year from asbestos-related diseases, mainly cancer [50, 51].

7.5.3.2 Sealants

As briefly discussed previously, PCB, a family of highly toxic and oily, non-flammable industrial chemicals, can exist at high levels in and around some old buildings due to the sealants based on polysulfide polymers containing PCB. PCB in many polysulfide sealants have been replaced now by chlorinated paraffins, and smaller volumes of chloroparaffins are used as plasticising agents for various sealants. However, the effects of chloroparaffins on health, is also a controversial issue, and its use in sealants is in decline. In sealants, especially in polysulfide and PU sealants, the BBP is also used.

Polyurea seals are 100% solid, with high elongation self-levelling elastomers. They are completely VOC free, used in horizontal saw or preformed joints on concrete or asphalt.

7.5.3.3 Paints, Varnishes and Lacquers

Paints, varnishes and lacquers contain various solvents as carriers, and when a paint is applied, a volatile component (mineral spirits and white spirits for alkyds, water for latex emulsions, alcohol for shellac and lacquer thinner for lacquers, in addition to xylene and naphtha, used as thinners) evaporates and the non-volatile portion of the paint (binders: liquid adhesives that form the surface film) is left on the surface. Binders can be natural (shellac, rosin, linseed oil) or synthetic (alkyd, epoxy, urethane and styrene-butadiene in water-based paints, as well as acrylic/vinyl acrylic latexes, latex referring to water-based). To comply with the VOC emissions laws, more solids are added which eventually makes the paint thicker and takes longer to dry. Depending on the drying speed, application and the type of finish that results, variations in the type of solvent used is possible. The most common solvents encountered in construction are: white spirit, xylene and 1-butanol. Some paints, long after their curing and drying, can still give off some residual odours, which can be due to a number of factors (i.e., due to unreacted monomers used in the manufacture of the resins and plasticisers or trapped solvents in small amounts).

The Clean Air Act demands the elimination of organic solvents use, hence a number of plasticisers and solvents (such as methylene dichloride (MEK), toluene, xylene) are classified as hazardous, which led to development of new, environment compliant systems to develop, such as 'water borne' systems , or use of rapidly curable/dryable oligomers (i.e., printing inks use oligomeric bis acrylates, polyester acrylates and polyurethane acrylates, which can be crosslinked quickly by UV light with the aid of certain sensitisers).

For water-based or latex emulsion type paints, there would be unreacted (free to evaporate) monomers, glycols and glycol ethers, alcohols, amines, and possibly formaldehyde, free monomers and plasticisers, that will continue to generate VOC for a long time after the paint has dried. However, the VOC level in a water-based paint is generally much lower than other common solvent-based products due to slower evaporation of VOC producing constituents. Glycols, such as ethylene and propylene glycol are very commonly used in interior and exterior latex/emulsion base coatings with concentrations of 2 to 5%, by weight, which evaporate very slowly. The glycol ethers are needed for proper film formation, they evaporate slowly and can remain in the applied coating film for about 72 hours or longer. Preservatives used in water-based paints were a mercury compound such as phenyl mercuric acetate or formaldehyde, and they are replaced by other organic preservatives now. Amines can be of different types: from fast evaporating (ammonium hydroxide) to slow evaporating (amino-2-methyl propanol). The types of free monomers in water-based/latex paints depend on the type of polymer (or its blend) used, and are the basic building blocks of the adhesive film forming polymer. Petroleum-solvent-based paints contain aliphatic and aromatic hydrocarbons, ketoximes, alcohols, free monomers and plasticisers. In a general purpose petroleum-solvent-based paint the volatile components are aliphatic hydrocarbons and paraffin naphtha. Odourless solvents (a isoparaffinic naphtha or petroleum distillate) used in some interior oil/alkyd-based coatings can pose a danger for hypersensitive persons. Fast drying paints generally contain certain aromatic hydrocarbons (toluene and butyl acetate) and smaller volumes of chloroparaffins are used as plasticising agents for various paints, however, the effects of chloroparaffins on health is still a controversial issue and its use in paint applications are in decline.

Varnish is a transparent (or pigment-less) film applied to stained or unstained wood, and contains volatile solvents such as white spirits and 1-butanol.

7.5.3.4 Adhesives, Polishes and Other Maintenance Materials

Solvent-based adhesives (commonly used on laminates, tiles, parquet and vinyl flooring) can contain alcohols, ketones, hydrocarbons, plasticisers, and some monomers.

Water-based adhesives can contain formaldehyde (as preservative), amines, glycol ethers, alcohols, plasticisers, and some free monomer (depending on the type of polymer and polymerisation used). Smaller volumes of chloroparaffins are used as plasticising agents for various adhesives.

With both types of adhesives, a long-term, slow release of the VOC are expected, due to their possible absorption into the surfaces and hence, sealing of the adhesive. Adhesives used in carpet backings can contain residual formaldehyde or isocyanates and vinyl acetate and various hydrocarbon solvents (from treatments and adhesives used to laminate the backing). From new carpeting, 4-phenylcyclohexane (a byproduct of styrene-butadiene commonly used to bind the backing) is usually emitted which can cause headaches, sore throat, lethargy, skin and eye irritation even at very low concentrations (1 ppb). Chlorinated paraffins (mainly CPVC) are widely used in PVC for adhesive applications, mainly to give greater resistance to ignition and combustion than general-purpose plasticisers. However, the effects of chloroparaffins on health, is still a controversial issue and its use is decreasing [23].

Cleaners, waxes and polish can contain a number of chemicals, beginning with a volatile carrier.

7.5.3.5 Wall-Coverings, Wall Papers

Although the terms wall coverings and wallpapers are used interchangeably, they may be made of paper backed by cotton fabric, vinyl faced with paper or cotton backing, fabric with a paper backing or may be all paper. Wall-coverings are mainly one of the flexible PVC applications (in vinyl coated paper, paper backed vinyl/solid sheet vinyl, fabric backed vinyl), and they may contain some free monomers (i.e., vinyl acetate, styrene, acrylic and even vinyl chloride), plasticisers (general purpose, mostly phthalates, such as DOP, DINP and DIDP are used), adhesives and certain preservatives. In general adhesives used for wall-coverings are modified starch products and special adhesives can be either styrene-butadiene or vinylacetate-ethylene copolymers with organic solvents and preservatives.

Chlorinated paraffins (mainly CPVC) are widely used in PVC for wallpaper applications, to introduce greater resistance to ignition and combustion than general purpose plasticisers. Hovewer, the effects of chloroparaffins on health is still a controversial issue and its use as flame retarders are decreasing [23].

7.5.3.6 Wires and Cables

Cable application of plastics is mainly done with PVC in the form of PVC-P, because of the economy and fire safety provided by the base polymer PVC. Chlorinated paraffins (mainly CPVC) is widely used in PVC to give greater resistance to ignition and combustion than general purpose plasticisers where CPVC are big size molecules with almost no emission probability. However, the effects of chloroparaffins on health is still a controversial issue and its use as flame retarders in PVC applications for cables are decreasing [23]. The most common plasticisers used in cables in Europe are DOP and DIDP. In high temperature PVC cable applications, mainly trimellitate plasticisers (preferably high molecular weight versions) are mainly used, without any health hazards. In insulation of wires and cables, EPS and PE, in addition to PVC are widely used.

7.5.3.7 Piping and Fittings for Plumbing and for Pressure Applications

Plastic pipe and fittings, particularly PVC-P and to some extent PE, are overwhelmingly being used in construction as a big competitor of metallics. APME's report [52] demonstrates that plastics' use in gas, sewage and water piping has tripled in the EU. Health hazards in PVC-P pipes and fittings, are subject to emission of the plasticisers used. Polyacetals are also used for sanitary ware and plumbing.

Pressure pipes (used in pipelines for gas, petrochemical and potable water applications in public utilities) are prepared from thermoset composite structures in general [53].

7.5.3.8 Flooring and Floor Tiles

Safe Kids (Washington, DC, USA), a non-governmental organisation, selected a playground material made from PVC, because they are found to be safe, non toxic and ideal in preventing the accumulation of allergic bodies such as fungi and mildew. In fact, PVC floors are commonly used in 'many nursing homes and hospitals, in particular in surgical theatres'. Since flexible PVC is needed in such flooring, it is necessary to use PVC-P. Usually general-purpose plasticisers (various phthalates, DOP, DINP and DIDP) are being used for this purpose. APME's report [52] demonstrates that plastics' use in floor coverings in construction increased considerably in recent years. Chlorinated paraffins (mainly CPVC) are widely used in PVC to have greater resistance to ignition and combustion than general-purpose plasticisers. However, the effects of chloroparaffins on health, is still a controversial issue and its use as flame retarders in PVC applications for flooring are decreasing [23]. For cushion vinyl flooring applications, several phthalates such as DOP are commonly used due to its economy and safety; BBP or DIHP, as well as plastisols of PVC plasticised with 2,2,4-trimethylpentan-1,3-diol di-isobutyrate are also used.

In addition to the common applications of PVC in flooring and floor tiles, there are also applications of PU (coatings for flooring), polyamide (PA) (woven carpets and rugs), linoleum (manufactured by oxidising linseed oil mixed with pine resin and wood flour in the form of sheets on jute backing), wood/thermoset composites (thermosets like PF, UF, polyisocyanates used as veneers, in plywood, particleboard, fibreboard, laminates), wood fibre/thermoplastic composites (for industrial flooring with PE).

7.5.3.9 Window Frames and Doors, Elements for Decoration and Furniture, Glazing and Decking, Repairing and Rehabilitation

The following plastics and rubbers are used preferentially for the following applications:

a) EPS, unplasticised PVC (UPVC), and thermoplastics composites with wood fibre (or flour) applications in window frames and doors.

b)   Mainly EPS, PU, acrylic cast sheets, polyacetals and PA in furniture fittings, wood/PF, UF thermoset composites, polyisocyanates (as veneers), in plywood, particle board, fibreboard, laminates, wood fibre or flour/PP, PVC, PE, PS, thermoplastic composites as elements for decoration and furniture.

c)   Acrylics, mostly PMMA and PC in glazing (in doors, windows, skylights and lighting devices), and PVC, as glazing.

d)   Wood fibre or flour composites with thermoplastics - such as PP, PVC, PE, PS, and thermosets such as PF, UF, polyisocyanates, in decking.

e)   Epoxies, glass fibre reinforced plastics (GFRP), carbon fibre reinforced plastics (CFRP), PU in repairing/rehabilitation.

7.5.3.10 Insulation Applications

For insulation applications, plastics are mostly used. For water insulation (waterproofing), PU, PVC, thermoplastic elastomers, silicones, various rubbers, acrylics, and even PE are mainly used. For heat insulation applications, PVC, EPS, PU foams are used directly (for roofs and walls), as well as in sandwich panels. In addition, UPVC-hollow panels are being used as 'interior cladding' to provide multi-purpose insulation (for water, heat, noise), specifically for pipes, wall insulation, and so on.

Cavity insulation applications, where a solid foam is produced *in situ* inside the cavity walls, without dismantling the walls; are done by direct injection of the liquid mixture of components and catalyst together into the cavity. This method is simple enough to carryout, but has the risks of carrying toxic chemicals and remnants of monomer with the foam structure, that can later migrate slowly indoors.

7.5.3.11 Structural Applications

Thermoset composites (GFRP, CFRP) are mainly used for load bearing applications in construction. Many load bearing foundations in buildings use rubber.

Most of this chapter is taken from the Chapter 12 (Construction Materials and Health Issues Indoors) written by Guneri Akovali, in the book 'Polymers in Construction', Ed., G. Akovali, Rapra Technology, Shawbury, Shropshire, UK, 2005.

The Author thanks Smithers Rapra Technology for their kind permission to reproduce it here.

References

1. D.A. Middaugh, S.M. Pinney and D.H. Linz, *Journal of Occupational Medicine*, 1992, **34**, 12, 1197.

2. *Clausen v Standard Insurance Co.*, 1997, U. S. Dist. Lexis 5873 (D. Colo., April 29, 1997) (D.C. Super. Ct. Civil Action No. 90-CA-10594, November 29, 1995).

3. *Weekly v The Industrial Commission*, 1993, 615 N E. 2d 59 (App. III, 1993).

4. J. Thornburg, D.S. Ensor, C.E. Rode, P.A. Lawless, L.E. Sparks and R.B. Mosley, *Aerosol Science and Technology*, 2001, **34**, 3, 284.

5. J. R. Riggs, *Materials and Components of Interior Architecture*, 6th Edition, Prentice Hall, NJ, USA, 2002.

6. V. Jowaheer and A.H. Subratty, *International Journal of Environmental Health Research*, 2003, **13**, 1, 71.

7. M.T. Bomberg and J.W. Lstiburek, *Journal of Thermal Insulation*, 1998, **21**, 4, 385.

8. *Alternative Construction: Contemporary Natural Building Methods*, Eds., E. Lynne and C. Adams, John Wiley & Sons, New York, NY, USA, 2000.

9. S.J. Reynolds, D.W. Black, S.S. Borin, G. Breuer, L.F. Burmeister, L.J. Fuortes, T.F. Smith, M.A. Stein, P. Subramanian, P.S. Thorne and P. Whitten, *Applied Occupational and Environmental Hygiene*, 2001, **16**, 11, 1065.

10. Y. Soma, H. Sone, A. Takahagi, K. Onisawa, T. Ueda and S. Kobayashi, *Journal of Risk Research*, 2002, **5**, 2, 105.

11. T. Tuomi, B. Engström, R. Niemela, J. Shinhufvud and K. Reijula, *Applied Occupational and Environmental Hygiene*, 2000, **15**, 8, 629.

12. D.J. Paustenbach, *Journal of Toxicology and Environmental Health. Part B: Critical Reviews*, 2000, **3**, 3, 179.

13. J.A. Tshudy, *Journal of Vinyl and Additive Technology*, 1995, **1**, 3, 155.

14. M.C. Robe, J. Brenot, J.P. Gambart, G. Ielsch, D. Haristoy, V. Labed, A. Beneito and A. Thoreux, *Indoor and Built Environment*, 2001, **10**, 5, 325.

15. M. Antonietta and S. Henke, *Chemie in Unserer Zeit*, 1995, **29**, 5, 275.

16. *Building Research and Information*, 1995, **23**, 6, 306.

17. R. Waring, *Private Communication*.

18. E.F. Group, *Journal of Vinyl and Additive Technology*, 1984, **6**, 1, 28.

19. *Opinion on the toxicological characteristics and risks of certain citrates and adipates used as a substitute for phthalates as plasticisers in certain soft PVC products*, CSTEE, EU, Brussels, Belgium.

20. *World*, 1984, **189**, 5, 60.

21. R.D. Deanin and Z.B. Zhang, *Journal of Vinyl and Additive Technology*, 1984, **6**, 1, 18.

22. C. Gabriele, *Modern Plastics International*, 1998, **28**, 9, 88.

23. A.S. Wilson, *Plasticisers - Selection, Applications and Implications*, 1996, Rapra Review Report No.88, Rapra Technology, Shawbury, Shrewsbury, **8**, 4.

24. *ENDS Report*, 1997, **266**, 11.

25. *ENDS Report*, 1997, **246**, 35.

26. L.S. Burn and A.P. Sullivan, *Journal of Water Supply: Research and Technology-Aqua Online*, London, UK, 1993, **42**, 3, 135.

27. L.S. Burn, and B.L. Schafer, *The Environmental Impact of Lead Leaching From UPVC Sewerage Waste and Vent Pipes*, CSIRO Building, Construction and Engineering Technical Report No.TR97/1, Highett, Victoria, Australia, 1997.

28. *Lead and You*, HSE, Caerphilly, Wales, UK, 2003.

29. *defra*, www.defra.gov.uk/environment/radioactivity/radon/index.htm

30. C.J. Howick and S.A. McCarty, *Journal of Vinyl and Additive Technology*, 1996, **2**, 2, 132.

31. C.J. Howick in *Proceedings of PVC'99*, Brighton, UK, 1999, p.233.

32. J. Williamson and B. Kavanagh, *American Industrial Hygiene Association Journal*, 1987, **48**, 5, 432.

33. B. Froneberg, L. Johnson and P.J. Landrigan, *British Journal of Industrial Medicine*, 1982, **39**, 3, 239.

34. P. Coghlan, *A Discussion of Some of the Scientific Issues Concerning the Use of PVC*, CSIRO Molecular Science, Clayton, Victoria, Australia, 2001.

35. C. Maltoni, G. Lefemine, A. Ciliberti, G. Cotti and D. Carretti in *Archives of Research on Industrial Carcinogenesis*, Volume II, Princeton Scientific Publishers, Princetown, NJ, USA, 1984.

36. *Plastics News*, 2000, **12**, 8, 19.

37. K.W. Shuh in *Handbook of Polymeric Foams and Foam Technology*, Eds., D. Klempner and K.C. Frisch, Hanser, Munich, Germany, 1991.

38. P. Keith, *Urethane Technology*, 1987, 4, 2, 38.

39. M. Judgeford, *Model Specification for Industrial Polymer Mortar Surfacings on Concrete Basis*, Building Research Association of New Zealand, Porirua, NZ, 1983.

40. A. Halstrom in *Proceedings of European committee for External Quality Assessment Programmes Laboratory Medicine (EQALM) Conference*, 2000, Boras, Sweden, Paper No.2.

41. *Product Standard*, Minnesota Statutes, Section 144.495, Department of Health, Minnesota, USA, 1985.

42. *Toxic Woods*, HSE, Caerphilly, Wales, UK, 2003.

43. A. Barry, *Measurement of VOCs emitted from particleboard and MDF panel products supplied by CPA mills*, Project Report No. 388N871, Canadian Particleboard Association, Canada, 1995.

44. M.D. Koontz and M.L. Hoang in *Measuring and Controlling Volatile Organic Compound and Particulate Emissions from Wood Processing Operations and Wood-Based Products*, 1995, The Forest Products Society, Proceeding No. 7301, WI, USA, 76.

45. T. Salthammer, *Journal of Coatings Technology*, 1996, **68**, 856, 41.

46. D. Franke, C. Northeim and M. Black, *Journal of Textile Institute*, 1994, **85**, 4, 496.

47. R.D. Leaversuch, *Modern Plastics International*, 2000, **30**, 12, 62.

48. M. Shubert, N. Mundinger and F. Hannes, inventors; Fritz Egger GmbH and Company, assignee; WO 5033204 A1, 2005.

49. *Asbestos Alert: Booklet of Health and Safety Executive,* HSE, Caerphilly, Wales, 2002.

50. *Asbestos Dust: the Hidden Killer,* HSE, Caerphilly, Wales, UK, 2003.

52. *Managing Asbestos in Work Place Buildings,* HSE, Caerphilly, Wales, UK, 2004.

52. *Plastics a material of choice in building and construction - plastics consumption and recovery in Western Europe 1995,* AMPE, London, UK, 1998.

53. T. Stafford, *Plastics in Pressure Pipes,* 1998, Rapra Review Report, No.102, Rapra Technology, Shawbury, Shrewsbury, **9**, 6.

54. *Biological Effects of Ionising Radiation, (BEIR) VI Report: The Health Effects of Exposure to Indoor Radon,* 1998, US Environmental Protection Agency, Washington DC, USA.

55. B.M.R. Green, J.C.H. Miles, E.J. Bradley and D.M. Rees, *Radon Atlas of England and Wales,* National Radiation Protection Board, Didcot, UK, 2002.

56. C.Y.H. Chao, *Applied Occupational and Environmental Hygiene,* 1999, **14**, 12, 811.

57. *Adhesives and Sealants Newsletter,* 1988, 7, 11, 1.

58. *US EPA and Surgeon General Call for Radon Home Testing,* Environmental News, 1988, USEPA, Office of Public Affairs A-107.

59. D.G. Mose and G.W. Mushrush, *Energy Sources,* 1999, **21**, 8, 723.

60. H. Arvela, *Science of the Total Environment,* 2001, 272, 1, 169.

61. V.C. Titov, D.P. Lashkov, I.M. Khaykovich and D.A. Chernik, *Applied Radiation and Isotopes,* 1997, **48**, 7, 997.

62. *Plastics in Building Construction,* 1988, **13**, 1, 4.

63. C.H. Sloan, R.L. Minga and T.H. Williams, inventors; Eastman Chemical Company, assignee; US 5399603, 1995.

64. K.F. Lindsay, *Modern Plastics,* 1989, **66**, 4, 149.

65. K.R. Kistler and E.L. Cussler, *Chemical Engineering Research and Design,* 2002, 80, 1, 53.

66. *European Plastics News*, 1999, **26**, 5, 49.

Some Additional Related Bibliography

1. N.A. Ashford and C.S Miller, *Chemical Exposures: Low Levels and High Stakes*, 2nd Edition, John Wiley & Sons, New York, NY, USA, 1998.

2. R.M. Bauer, K.W. Greve and E.L. Besch, *Journal of Consulting and Clinical Psychology*, 1992, **60**, 2, 213.

3. G. Heady, *Texas Tech Law Review*, 1995, **26**, 3, 1041.

4. H. Hirsh, *Medical Trial Technique Quarterly*, 1996, **43**, 1.

5. R.W. Katz, and J.N. Portner, Environmental Law Trial, 1993, **29**, 38.

6. I. Kerstin, *New York University Environmental Law Journal*, 1999, **7**, 1, 119.

7. R. Menzies, R. Tamblyn, J-P. Farant, J. Hantey, F. Nunes and R. Tamblyn, *The New England Journal of Medicine*, 1993, **328**, 12, 821.

8. L.A. Morrow, *Official Journal of American Academy of Otolaryngology - Head and Neck Surgery*, 1992, **106**, 6, 649.

9. S. Pfeiffer, *Boston Globe*, 1999, March 17th, Section B, p.1.

10. A. Seidner, *Hospital Practice*, 1999, **34**, 4, 127.

11. G.H. Wan and C.S Li, *Archives of Environmental Health*, 1999, **54**, 1, 58.

12. M.G.D. Baumann, *Volatile Organic Chemical Emissions from Composite Wood Products: A Review*, USDA Report, USDA, Washington, DC,USA, 1997.

13. *Formaldehyde Product Standard,* Department of Health, 1985, Minnesota, Minnesota Statutes, Section 144.495, USA.

14. J. Emsley, *Nature's Building Blocks*, Oxford University Press, Oxford, 2002, 33.

15. D. Anink, J. Mak and C. Boonstra, *Handbook of Sustainable Building: An Environmental Preference Method for Selection of Materials for Use in Construction and Refurbishment,* 2nd Edition, James & James Science Publishers, London, UK, 1998.

16. *Towards More Sustainable Construction: Green Guide for Managers on the Government Estate*, DETR, London, UK, 1999.

17. *Sick Building Syndrome: Concept, Issues and Practice,* Ed., J. Rostron, E & F. Spon, London, UK, 1997.

18. P.J. Jackman, *Specification of Indoor Environmental Performance of Buildings,* Building Services Research & Information Association-BSRIA, Bracknell, UK, 1987.

19. *Handbook on the Toxicology of Metals,* 2nd Edition, Eds., L. Friberg, G.F Nordberg and V.B Vonk, Elsevier, Amsterdam, TheNetherlands, 1986.

20. M.A. Kamrin, *Toxicology: A Primer on Toxicology Principles and Applications,* Lewis Publishers, Chelsea, MI, USA, 1988.

21. *Japan Solid Wood Products - Sick House Syndrome in Japan - A Key Issue, 2001,* USDA-GAIN Report No. JA1054, http://ffas.usda.gov/gainfiles/200107/120681333.pdf

22. J. Bower and L.M. Bower, *The Healthy House Answer Book,* The Healthy House Institute Publishing, ID, USA, 1997.

23. S. Enviro, *Home Clean- Chemical-Free Building Materials Against SB,* 2001, www.pref.shiga.jp/event/messes-e_kokusai/03b.htm

24. T.A. Loomis and A.W. Hayes, *Loomis's Essentials Of Toxicology,* 4th edition, Academic Press, New York, NY, USA, 1996.

25. M.O. Amdur, J. Doull and C.D. Klaassen, *Casarett and Doull's Toxicology: The Basic Science Of Poisons,* 4th Edition, Pergamon Press, New York, NY, USA, 1999.

26. H. Fisch in *the Proceedings of Eurochem Conference,* Toulouse, France, 2002.

27. P.A. Toensmeier, *Modern Plastics International,* 2001, 31, 8, 11.

28. J.S. Amstock, *Handbook of Adhesives and Sealants in Construction,* McGraw-Hill, New York, NY, USA, 2001.

29. *Di(2-ethylhexyl) Phthalate,* IARC Monograph. International Agency for Research on Cancer, World Health Organisation (WHO), 2000, 77.

30. *Chinese Markets for Construction Plastics,* Asia Market Information and Development Company, 2002.

Appendix

A-7.1 Radon Indoors

Uranium-238 is present in small quantities in most of the rocks and soil with a half-life of 4.5 billion years, decaying to other elements of radium by itself then to radon-222. Radon-222 continues to decay with a half-life of 3.8 days, to radioactive elements of radon daughters [54], or progeny (polonium-218 and 214, both emit high energy, high mass, alpha particles), which are very effective at damaging lungs. Radon daughters attach to lung tissues by inhalation, which means a deposition of a radiation source is possible inside the lungs, and hence they begin to emit alpha particles to the lung tissues. Since alpha radiation cannot travel long distances, they do not reach other organs in the human body, and for this reason lung cancer is the most important cancer hazard produced selectively from this effect.

The National Academy of Sciences of US estimated that radon from soil causes 15-22,000 deaths from lung cancer each year in the US.

The EPA set a concentration of 4 pCi/l of air inside homes as an action level, the average radon level in homes is normally around 1.25 pCi/l. Although there is US Congress Legislation from 1988 establishing the goal not to exceed 0.2-0.7 pCi/l in homes, this has not been found to be practical because of radon levels in ambient air indoors.

One can also be exposed to radon by ingestion of underground water that is contaminated with radon.

Radon is the biggest possible contribution to radiation exposure in houses (50%) that occurs naturally [29], and it is a gas, that has no taste, smell or colour.

Radon is existing everywhere but it is usually in insignificant levels that pose negligible health risk normally. It is a radioactive decay product of uranium (over radium). Uranium is usually found in all soil and rocks in small quantities varying from place to place, and radon also exists everywhere in varying amounts.

Due to the effects of wind and temperature, the air pressure inside the house is usually lower than the air pressure in the soil beneath it and air containing radon from the soil creeps into lower pressure area of the house (through cracks and gaps in the floor or walls). When radon rises from the soil to the air, outdoors, it is diluted enough, however, when it enters enclosed spaces, high concentrations can be built up indoors with a serious risk to health. Especially in the case of air with already high levels of radon originally, indoors can rise up to very high dangerous levels easily.

Radon gas concentrations are measured in units called 'Becquerels per cubic metre Bq/m³' and a *level of 200 Bq/m³ (max) is considered as the action level for homes.* This value is double for offices, taking into account that more time is usually spent in the home than at work. The usual level of radon observed normally is 1/10th of the action level (20 Bq/m³). In a study done in UK, it was shown that levels of radon vary considerably from location to location with the possibility of reaching values well above the action level suggested (called 'high radon potential areas') [55].

One should realise that radon is not only a problem in basements, but it can exist even on the upper floors of high-rise buildings. In an experimental study, indoor radon levels were monitored continuously with and without air-conditioning in a number of high-rise office buildings in Hong Kong [56], and it was found that the average indoor radon level during office hours were not as low as expected from the high rise positions (they were between 87-296 Bq/m³ with ventilation, which was some 25% lower than without ventilation). The average radon emanation rates were found to vary between 0.0019-0.0033 Bq/m³ for different high rise buildings and it was estimated that building infiltration rate accounted for about 10-30% of the total building ventilation rate in the buildings depending on building tightness [56]. On the contrary, there are also reports claiming that sealing homes to save energy does not concentrate radon indoors [57].

Radon levels indoors can be measured by a safe and a rather simple method by use of detectors [58]. Detectors are usually in the living room and in another occupied bedroom. In one application (in UK), the detectors used were just a piece of spectacle lens plastic put in a protective shell, about the size and shape of a small door knob (obtainable through mail order in some countries [29] and were returned after three months of testing in a reply paid envelope provided). The plastic in this system records radon, which was measured by accredited laboratories after its return. There are also much shorter (i.e., fortnightly) measurements available, that can be used for screening purposes as they are less accurate [29].

Indoor radon measurements obtained for homes in North Virginia, USA, revealed that existing high or low median indoor radon levels in each house persist through four seasons [59], however, attempts to compare between the soil radon outside and the soil permeability was not successful.

There are several studies showing the main defects in design and implementation to avoid high levels of radon indoors and to give guidance on radon-safe buildings in slab-on grade houses [29, 60, 61]. It is certainly best to stop radon entering the house first, and if this is impractical, then effective removal (or dilution of it) is recommended. It is shown that there are several ways to achieve this [29, 60]. The prevention (or decrease) of the flow of radon-bearing air indoors can be done through installation of aluminised bitumen felt as well as by use of elastic sealants to seal cracks and gaps in solid concrete floors and walls. As a precaution, it is suggested that perforated piping be installed in the subsoil of the floor slab. There are a number of studies such as on sub-slab ventilation matting [62] as well as production of alpha particle radiation barriers of sulfopolyester acrylic copolymer [63] and Nylon/polyester matting [64], and others [65, 66].

Installation of a radon sump system equipped with a fan is suggested as the most effective and best choice for high levels of radon existance [29]. For this, a sump which is a small empty space about the volume of a bucket is dug under the solid floor and a pipe is routed from it to the outside air. The sump and the fan connected at the exit of the pipe to suck air, both help to alter the air pressure below the floor and to pass and to release it harmlessly into the atmosphere. There are also applications where the fan is replaced by a blowing system to facilitate the removal of the remaining of radon in the soil. It is also possible to increase the circulation of air beneath the floor (improved ventilation under suspended timber floors with or without a fan *via* an air-brick) or at loft level or even by using positive house ventilation as a whole ('positive pressurisation' is most effective if house is very air-tight).

All of these are common methods that have been suggested and applied.

8 Use of Plastic and Rubber in Various Applications and Possible Health Effects

8.1 Plastic and Rubber Use in Sports and Leisure and Possible Health Effects

It is rather difficult to consider sports (such as soccer, golf, and athletics) and leisure (other activities, such as walking, sailing, horse riding) separately, however, there is considerable amount of plastic and rubber use in these sectors. Overall spending on sports goods and recreation for the US alone were given as about $40 billion in 1998, and this figure has kept rising at a pace of about 6% annually [1], which can be taken as a measure of plastics and rubber use in this sector. Certainly, the trend of prefering to use plastic and rubber arises mainly from the need for lightweight, economical and durable materials, which is generally provided by these materials, for playing surfaces to tennis rackets, footwear and clothing.

In 1998, polyurethanes (PU) and engineering plastics were shown to make up some 58% of advanced materials used in sporting goods [2], and for the moulded sports products, polypropylene (PP) is used three times more, compared to the next most used plastic, acrylonitrile-butadiene-styrene (ABS) [1]. On the other hand, there are also trials for use of other plastics, even the use of 'tyre crumb' for artificial surfaces, carpet underlays and playgrounds, although the use of the latter is not well developed yet (amounting to some 10% of the tyres crumbed).

Several examples are given next for some of these applications, along with the types of polymers used and some health effects.

8.1.1 Plastics and Rubbers as 'Artificial Surfaces' in Sports and Leisure

The first 'synthetic grass pitch' was applied in 1966 at the Houston Astrodome [3], and after that, rubber-based athletic tracks have continued to be used since the 1970s. Since then, both indoor and outdoor synthetic playing surfaces have become very common.

For artificial surfaces used in sports and leisure, one can note the following:

a) *Synthetic grass surfaces* are mainly composed of a base (a foam shock pad layer) with a synthetic carpet on top of it. They are designed mainly for outdoor use, hence their sustainabilities are very important. The base is usually made of rubber, or from flexibilised (heavily plasticised) plastics, and there are cases where even waste materials are used. The type of material and plasticiser used in the base are two important parameters that characterise whether synthetic grass has any potential health problems or not.

The upper part of synthetic grass is usually made from polyamide (PA), or polyolefin mixtures such as a mixture of polyethylene (PE) and PP (FieldTurf) [4]) or polyester (saturated), and certainly the type and amount of the flexibiliser used are important as regards the health issues of any of these.

Synthetic grass pitches are currently banned for use at higher levels (i.e., as soccer grounds).

b) *Sports hall floors* are an another area of use, where polymers are sheets or carpet usually overlayed above a wooden floor, to provide the proper degree of spring. For this, rubber, plasticised polyvinyl chloride (PVC), PA or PP are used.

Carpets prepared on PU foam backing are used for court games i.e., tennis.

c) *Athletic tracks* and *playgrounds for children,* are usually rubber-based. It can be a solid-cast rubber (mainly of PU or natural rubber (NR) or polychloroprene latex, as well as sheets of styrene-butadiene rubber (SBR), or it can be a rubbery sheet (i.e., plasticised PVC sheet), or a resin bound rubber crumb, or even a laminate composite structures of these. One such track was successfully used first in the Olympics in 1964.

e) *Synthetic snow and synthetic ice* is an another application of polymers, where artificial surfaces are produced. As an example, a high-density polyethylene material, in the form of a coated wooden panel, is shown to have at least 90% of the glide factor of ice and has a minimum lifetime of years (called super ice) [5].

In general, silica filled NR tiles are used for their high wear resistances, on the floor in skating rinks.

On synthetic ski slopes, mainly polyolefin units along with PVC and PA brushes are applied in a special arrangement.

In all of these applications, the type of plastics/rubbers used and the type of additives they are processed with, are important in deciding whether there is any health hazard existing and the level of it.

8.1.2 Plastic and Rubber Use as 'Clothing' in Sports and Leisure

In all sports activities, special clothing (i.e., breathable fabrics such as GoreTex for walking and climbing, and protective clothing) is necessary, which can be made of synthetic fibres. A different application is the preparation of fire resistant materials for motorsports, and footwear (where a variety of polymers such as rubber outsole and cellular vinyl acetate or PU midsoles with torsion bars made from carbon fibre reinforced polyester, along with uppers from PVC coated fabrics). Special shoes for soccer players and running shoes (with semi-rigid soles of thermoplastic PU), and boots for skiing and skating (thermoplastic shell with a number of other thermoplastic attachments), are also mainly composed of plastic and rubbery components.

8.1.3 Plastics and Rubbers Use in 'Water and Motor Sports'

'Economy of the material' and 'improved strength-to-weight ratios' are two strong points that made polymers distinct in their use in water and motor sports.

Boats and other water sports equipment made of 'carbon/aramid/glass fibre reinforced plastic materials and 'rigid foams' with a number of different design concepts have been constructed. As an example, boats with strong sandwich constructions generally have a foam core between reinforced composite panels, namely, 'carbon fibre reinforced epoxy skins and a PVC foam core structure', and even a catamaran with a 'thermoformed skin, a cellular core and a glass reinforcing layer' [6], has been constructed. In most of these, there are reinforced thermoset matrices in common, however, there are also cases where thermoplastics are used (as in the case of the multipurpose rowing/sailing boat moulded from PP, and a pedal-boat from rotomoulded PE [7, 8]).

In motor sports, since the weight of the vehicle is of prime importance, composite thermoset and thermoplastic dominated applications have been used for the last 40 years, while rubber is an another important area of application (beginning from large tyres, to seals, bushes, bearings such as self-lubricating polymer bearings and various mountings). In addition, a number of other polymeric materials such as plasticised PVC, PU and their foams are extensively used in the interior trim and seating in racing cars.

In this category, because these compounds are used outdoors, the expected health hazard will be not from volatile organic compounds (VOC), but mostly from the flammability of the material.

8.2 Automotive and Transportation Applications

The automotive sector is one of the three main markets for plastics, with a share of 15% of total polymers produced [after packaging (40%), building and civil engineering

(25%)], and it is the first for rubbers.

This consumption overall corresponds to about 7 million tonnes of the total polymers produced. It is estimated that, more than 500 million passenger cars are currently in use, and 50 million new cars are being produced each year, worldwide.

Fogging which is emission of additives from plastic and rubber materials used in the interiors of the car is a common problem, which is most obvious when a car is new. Fogging is simply the condensation of VOC (emitted in car interiors) on colder surfaces, particularly on the windscreen, leading to a 'clouding' on the glass surfaces. It is known that evaporation of plasticisers from dashboard materials contribute greatly to the fogging, but phosphorus flame retardants used in flexible PU foams under the seats also have a considerable contribution (when the flame retardant used is volatile or has volatile impurities).

8.2.1 Why Use Plastics and Rubbers in Automotive Applications?

There are several reasons why plastics and rubbers are used in automotive applications:

a) Some of their properties are unattainable by the conventional materials i.e., damping (hence the use of the noise reduction properties of PU, PE, PVC and PP and their foams), elasticity, low moduli and sealing properties of rubbers and thermoplastic elastomers (TPE).

b) They have an exceptional combination of properties which are unattainable from other materials i.e., metals and glass.

c) They are lower in weight for the same volume of material (a very important fact, because carmakers are aiming to lighten the vehicles to reduce fuel consumption and emission of pollutants (every 10% reduction in mass is shown to reduce fuel consumption about 6% [9]).

d) They are economical on a volume basis (they offer an economical response to mass production).

e) they allow much more versatility in design freedom and better mechanical responses within limits (consider rubber with better resilience and shock absorbing characteristics), ready-to-use parts can be easily produced and applied.

f) They are non rusting.

g) They are widely open to new improvements and developments (i.e., application of nanocomposites give much higher mechanical properties and thermal resistances, use of conductive polymers for direct painting, development of new hybrid plastic-metal technologies, such as the use of metal-polymer sandwich structures for car bodies [10]).

It is estimated that, over the past decade, about 300 kg of traditional materials have already been replaced during the construction of the cars by 100 kg of plastics, resulting in much lighter cars and a reduction in fuel consumption (amounting to 750 litres less fuel over the whole life of the car, which is about 12 MT of oil for EU, as well as reduction of related carbon dioxide emissions by 30 MT. Their low moduli and strengths, in general, as well as the negative public image about their use, are the main handicaps for the use of plastics and rubbers in the automotive industry at the moment.

8.2.2 Which Plastic/Rubber to Use for Automotive Applications?

Almost all of the polymer families are involved in automotive applications: there are thermoplastics, where automotive applications is the third market in its consumption (with 13%), there are thermosets, where automotive is the second market in their consumption (with 14%), there are composites, a series of rubbers, and (TPE), for the latter three, automotive is the first market in their consumptions.

Some plastics are used on almost all vehicles produced (such as PP, PA, PE, ABS or PU), while others are used only on specific models (such as the carbon-reinforced composites in racing cars and others, i.e., the Mercedes SLR Maclaren has a body and chassis with front crash structure made entirely of carbon fibre composite with comparable strengths, but with a 50% reduction in weight as compared to the same parts in steel). While some of them are used in high quantities for large parts, others are only used for specific parts in smaller quantities (such as fluorinated hydrocarbons). PA, for example, are used preferably for under the hood applications because in a car, engines are running at high temperatures, requiring plastics that are heat resistant, especially in the presence of harsh automotive fluids.

Thermoplastic matrices, is particularly attractive for automotive applications: PP, for example, is economical, it can be processed quickly and can provide much better mechanical properties such as impact resistance, (i.e., bumpers, body panels [11]). PA are successfully applied in both under hood (i.e., inlet manifolds, radiator fans) and interiors (instrument panels, doors, front-end structures). For better temperature performance and mechanical properties, in some special application areas (motor racing sector, gearbox parts), polyether ether ketone and polyphenylene sulfide ('high performance thermoplastics') are also used as matrices.

There are plastics composite applications for automotive bodies, bumpers, etc., and nanocomposites are already used in some interior parts (i.e., a structural seat back of a Honda car, is already made from a PP composite having 6% nanoclay instead of the usual

30% glass fibre reinforced version of it). Increase of composites use in various parts of the cars, beginning from car bodies to the chassis, powertrain and fuel tank, is under study.

In this respect, (thermoset) plastics composites with discontinuous fibre products are already mostly used in the car body applications, where polyester/E-glass is predominating (mostly because of polyesters, economy, ease of processability and reasonable mechanical properties provided), followed by use of phenolics (when fire retardance is required, in friction linings and engine compartments), and epoxies. Replacement by carbon or aramid fibre reinforcements can reduce body mass by 40% (compared to steel) and with more added strength, but the cost is unfavourable at the moment, as mentioned previously [12, 13].

The types of rubber for the automotive industry can be considered in two groups: tyres (mostly of NR, SBR, polybutadiene and butyl type rubbers), and profiles and seals (for profiles; nitrile-butadiene rubber and derivatives, acrylates, fluoro-rubbers; and silicones for sealing).

Plastics, rubbers and composites are mainly used in automotive applications as damping and protective devices (i.e., PU and other foams, are used for safety cushioning and acoustic, thermal and water insulation in seats, and in airbags for shock absorption), for sealing (from liquids, dust, rain, air), they are used in the body (as external elements and structural load bearing parts), and in the passenger compartment (which consumes more than 50% of the plastics used in a car), as well as their applications under the hood, and as glazing (as an emerging application).

PU are mainly used as structural foams (as a core component and interior trim panels), as well as non-structural components (instrument and door panels, seat-back and head liners, and as foam in seats).

All of the plastics, rubbers and composites described previously for their use in the automobile industry may or may not show any health hazard, depending on their preparation and contents. In this context, certainly the ones that are used in interiors are of the utmost importance and should be considered and examined more critically. On the other hand, the ones used in exteriors, should also be considered, not for their health hazard considerations, but mainly on different grounds (i.e., the loss of light stabilisers by migration from automotive finishes, which can significantly reduce the durability of coatings on plastic substrates [14]).

Composite plastics and rubbers are being used preferably in other transportation applications (wagons of metros, trains, airplane interiors), where flammability as well as VOC are important. In general, PF or epoxy composites are used, and certainly it is compulsory to use carefully selected flame retarders in such mass-transport systems. In such applications, it should always be remembered that, not only flames are important in these applications, but, the smoke production is very important and critical too.

8.3 Plastic Use in Agriculture and Possible Health Effects

The applications of plastics in agriculture are rather limited: in western countries its share is about 4%, and a bit higher in the agriculture-intense countries (above 10%). There are a number of important applications of plastics in agriculture, such as, ground covers, which are used to conserve moisture to raise the soil temperature and to prevent nutrient loss and to inhibit interaction of weeds and insects, which can increase the crop yield by 200 to 300%. Another example to consider is their use as containers, binders and twines and as netting and as temporary covers for storage of the harvest. Application of plastics as greenhouse covers and as mulching films are the most common. Plasticulture is the word used for them. However, use and selection of plastics must still be done very carefully because in most cases plastic comes in contact with soil for a long time, or can even be mixed with the soil at the end of use in the case of the use of biodegradables. Any excess remnants of heavy metal ions can accumulate in the soil, affecting the crops and underground water sources.

8.4 Plastic and Rubber in Electric and Electronics Applications, Their Health Effects

Developments in polymers for electronic applications e.g., conductive and intrinsically conducting polymers, new polymer dielectrics and electro-active polymers, photoresists, encapsulants, underfills and adhesives, polymers in displays, active devices and waveguides, all show the importance and versatility in the use of plastics. The ever increasing need for faster microelectronic devices with higher memory capacities led to increase in number of devices per chip, along with the need for sub-micron structures (photo resist technology, etching) to decrease device dimensions are best coped with by selection and use of proper plastics materials. A large variety of polymers, such as polyimides, polyarylene ethers, polybenzoxazoles, SiLK (a new generation class of ring-structured polymers with very low dielectric constant (2.6-2.7) and high decomposition temperatures (500 °C), developed by Dow), and polynorbornanes, are good candidates for low dielectric constant applications in microelectronics [15]. Electricity and electronics need a broad range of properties, such as rigidity (or flexibility), electrical insulation or conductivity. It is estimated that about 7-8% of total plastics are being used in this sector, being mostly thermoplastics (with 76% share) and thermosets and composites (24% share) [16].

One recent feature of polymers in electronics use is 'conducting polymers'. Polymers are known as intrinsically non-conducting or insulating electrically. This is an advantage in most cases, but in some others, we want to have a plastic or rubber which conducts electricity, better as good as or better than copper. Until 1977, polymers were made conductive by blending them with a conducting material, such as a metal or carbon powder. However, in 1977, after the report of the first intrinsic conducting organic polymer (doped acetylene), although these new generation conducting polymers, called 'intrinsically conducting polymers' were neither processable nor air stable, there has been a lot of interest in them and a new era was born [17].

Later, these intrinsically conducting polymers were improved and a new generation of processable, and air stable conducting (thermoplastic) polymers were introduced, and found immediate applications in electronics industry as 'electrodissipating packages', [18].

Intrinsically conducting polymers are very promising materials for shielding electromagnetic radiation and for reducing or eliminating electromagnetic interference. Certainly, they have the advantage of being polymers as well: they are light weight, non-corroding and flexible, which already gave rise to a number of military and commercial applications, such as their use in light emitting polymer devices (by incorporation of conducting polymer fibres); and a widespread use of polymers in new devices used for sensing, chemical separations, electromechanical actuators, electronic components and electrochromics [19, 20].

In a recent development, an 'electronic packaging' system was introduced to provide electrostatic shielding and moisture barrier desiccating package with air cushioning.

Wire and cable applications of plastics and rubbers are apparent from wirings used in buildings and automobiles, as well as energy cables. For the EU, the wire and cable market is estimated at roughly €18 billion. PE is the most consumed polymer for wire and cable applications, both as a thermoplastic, crosslinked thermoset and foamed PE; while PVC is the second, followed by the others (rubbers, PP, TPE, PA, PU, silicone, and fluoropolymers) [16]. Long-term durabilities are usually a must for them and the uppermost covering layers specifically are expected to be resistant to UV or ozone ageing in general, resistant to moisture or water for underwater cables, and oil resistant for industrial applications, which can be maintained by use of proper active functional additives as well as by selecting of proper plastics for the specific application.

Polydiene rubber is a common low voltage electrical insulation material, used because of its low electrical conductivity and elasticity. But, it can degrade rapidly by peroxidation, even if there are antioxidant stabilisers in the system, which can lead to electrical shortages and fires. Polydiene rubbers are recently being replaced by saturated ethylene-propylene (EP) rubbers or plasticised PVC.

There are certain stringent regulations put forward with the aim to help to reduce waste generated by old electrical and electronic equipment, which forced development of easy-to-recycle materials as well as halogen-free fire retardant solutions, such as bisphenol A diphenyl phosphate, in addition to the traditional resorcinol diphenyl phosphate.

8.5 Outline of Plastics Use as Other Consumer Products and Possible Health Effects

Consumer products is an another sector which uses plastics at large (i.e., in the USA alone, it is estimated that, more than 5,000 tons of plastics per year is used for consumer goods production). A number of different applications, like toys, nappy backings, eating utensils, bodies of watches and cameras, various hygiene articles, personal articles like combs and handles of razors, and household appliances, can be found in this category.

References

1. J. Sloan, *Injection Molding*, 1998, **6**, 10, 96.

2. A. Warmington, *European Plastics News*, 1999, **26**, 11, 33.

3. R.P. Brown, *Polymers in Sport*, Rapra Review Report No.135, Rapra Technology, Shawbury, Shrewsbury, UK, 2001, **12**, 3.

4. *Field Turf*, GSA, www.fieldturf.com

5. F. Esposito, *Plastics News (USA)*, 1997, **9**, 15, 28.

6. *Plastics Technology*, 1995, **41**, 11, 92.

7. *Plastics Engineering*, 1999, **55**, 6, 7.

8. L.S. Neaville, *Plastics News* (USA), 1996, **8**, 12, 4.

9. C.D. Rudd, *Composites for Automotive Applications*, Rapra Review Reports No.126, Rapra Technology, Shawbury, Shrewsbury, UK, 2000, **11**, 6.

10. *Advanced Materials Newsletter*, 1998, **20**, 6, 1.

11. W. Baumann, *Journal of Macromolecular Science A*, 1999, **36**, 11, 1743.

12. *Advanced Materials and Composites News*, 1998, **20**, 18, 3.

13. R. Voelz, *Kunstoffe Plast Europe*, 1996, **86**, 3, 39.

14. P.V. Yaneff, *JCT Research*, 2004, **1**, 3, 201.

15. S. Banerjee, *International Journal of Plastics Technology*, 2003, **6**, 102.

16. M. Biron, *Electricity and Electronics like Polymers I – Economic and Technical Outline*, Omnexus News, http://www.omnexus.com

17. C.K. Chiang, C.R. Fincher, Jr., Y.W. Park, A.J. Heeger, H. Shirikawa, E.J. Louis, S.C. Gau and A.G. MacDiarmid, *Physical Review Letters*, 1977, **39**, 17, 1098.

18. R. Kohlman, J. Joo and A.J. Epstein, *Physical Properties of Polymers Handbook*, Ed., J.E. Mark, AIP Press, New York, NY, USA, 1996.

19. *Conductive Polymers and Plastics in Industrial Applications*, Ed., L. Rupprecht, Plastics Design Library/William Andrew Publishing, Norwich, NY, USA, 1999.

20. A.K. Bakshi, *Bulletin of Materials Science*, 1995, **18**, 5, 46.

Some Additional Related Literature

1. M. Biron, *Thermosets and Composites: Technical Information for Plastics Users*, Elsevier, Oxford, UK, 2004.

2. F. Pardos in *Proceedings of the SPE ANTEC Conference 2002*, San Francisco, CA, USA, 2002, p.2736.

3. F. Szabo, *International Polymer Science and Technology*, 2001, **28**, 11, 1.

9 Sustainability Through Plastics and Rubbers

9.1 Sustainability in General

Sustainability is the 'ability to achieve economic prosperity, while protecting nature, and to provide a high quality of life for everyone'. Hence, 'sustainable development' is being responsible at every step of production (i.e., production, use and disposal of materials), by considering the triangle of 'environmental, economic and social' impacts. In this triangle, environment means waste and energy mainly, and society means goods and services, and the area where all three points overlap characterises any sustainable activity. The most widely accepted definition of sustainable development, is from the Brundtland Commission (UN Commission on Environment and Development), which is stated as 'the development which meets the needs of the current generation without jeopardising the needs of the future (generations)' [1, 2], which simply is concerned with 'pollution' and 'the reckless use of natural resources', and for plastics and rubbers, it also means 'waste minimisation' and 'reduction of non renewable raw materials use'. In both concepts of sustainability and sustainable development, there will be the 'avoidance of any toxics' concept in any of the plastics and rubber element involved. The product and process oriented, Life Cycle Assessment can be used as a tool for assessing the environmental performance of an activity during monitoring sustainable development [3, 4]. The efforts to achieve sustainable solutions and sustainable development in general calls for everyone to take responsibility to help in solving all related environmental problems, namely, all individuals, communities, businesses and governments.

Gandhi is the one who continuously rejected the Western model of development, who once said 'it took the exploitation of half the globe to make Britain what it is. How many globes would it take India to do the same?' It was later, that this was answered indirectly, by Brundtland, the Chair of the World Commission on Environment and Development, who stressed that humanity would need ten Earths to allow all humans living on Earth to reach to Western levels of consumption [5].

Whenever proper sustainability is achieved, certainly there will be more care for the pollution and toxics as well as the health related issues in connection with plastics and rubbers. Hence, sustainability is very important to reach an end of the possible health problems encountered with plastics and rubbers.

187

9.2 The EU - Sustainable Development Strategy (SDS)

SDS is the broad and long-term strategy in the EU. A review is undertaken every five years in the EU, of the sustainable development strategy in general, by exploiting the benefits of eco-innovation and developing sustainable products and processes for emerging global markets [6]. According to Eurostat statistics and to the strategy report, the production of toxic chemicals in the EU-15 has more or less stabilised, after having a peak in 2000. But, the real picture is a bit different: the production of the most dangerous group, namely carcinogenic, mutagenic and reprotoxic chemicals, has grown by 3.4% per year, and the least harmful category has shrunk by 4.3% per year since 2000 [7]).

The EU launched a strategy to reduce diseases linked to environmental factors, (environment and health [8]). Around 25-33% of diseases in industrialised countries are due to the environmental factors. The EU Commission will be launching pilot schemes on pollutants (with specific relevance to children), including dioxins, heavy metals and endocrine disrupter chemicals (ECD). Asthmatic symptoms (some 10%), allergies and other respiratory diseases are one of the main reasons for hospital treatment, and less common diseases are increasing (such as leukaemia, which is the most common form of childhood cancer). Many pollutants, including dioxins, are known to be too dangerous even before birth, because they can reach the foetus in the womb and increase the risk of miscarriage and birth defects. In some parts of Europe, it is also found that, about 10% of infants develop mental or physical disabilities, which is blamed on heavy metals in the environment, such as lead and mercury compounds. Chemicals, such as ECD, can also act like hormones and disrupt the way the body functions, with adverse effects on reproductive capacity. In large European cities, it is estimated that around 60,000 deaths per year are caused by long-term exposures to all sorts of toxics found in polluted air. The Commission announced in its Communication on Integrated Product Policy (COM (2003) 302 final) that it will seek to identify and stimulate action on products with the greatest potential for environmental improvement, to be carried out in three phases. The first phase, Environmental Impact of Products (EIPRO) aims at identifying the products with environmental impact from a life-cycle perspective, in the EU. Consumption has been grouped into different product categories. The part completed in May 2006 shows that products from three areas of consumption, (food/drink, private transportation, and housing) are responsible for 70-80% of environmental impacts. In the second phase Environmental Improvement of Products (IMPRO), possible ways to reduce the life-cycle environmental impacts of some of the products with the greatest environmental impact, will be identified. The research involved in phases one and two are led by the Institute for Prospective Technological Studies in Seville, Spain, and the IMPRO project will be carried out in 2006-2007. In the third phase after 2007, (which will have Policy implications), the EU Commission will seek to address policy measures for the products identified as having the greatest potential for environmental improvement with the least socio-economic cost.

There is also the EU Restriction of the Use of Certain Hazardous Substance Directive (RoHS) in electrical and electronic equipment which bans new electrical and electronic equipment with more than agreed levels of cadmium, hexavalent chromium, lead, mercury, polybrominated biphenyls (PBB) and polybrominated diphenyl ether (PBDE) flame retardants, in the EU. The RoHS Directive and the UK-RoHS regulations came into force on 1st July 2006 [9]. To prevent the generation of hazardous waste, the Directive 2002/95/EC requires the substitution of all heavy metals (cadmium, hexavalent chromium, lead and mercury) and brominated flame-retardants (PBB or PBDE) in new electrical and electronic equipment sold as of 1st July 2006 [10].

9.3 A Briefing on Environmental Laws and Sustainable Use of Plastics and Rubbers

Over recent years, much environmental legislation appeared (mostly imposed on manufacturing industry). Energy Using Products (EuP), Integrated Performance Primitives (IPP), RoHS and Waste Electrical and Electronic Equipment (WEEE), to name but a few that directly impact on the manufacturing sectors that produce or use plastics or rubbery materials in their products.

9.3.1 Plastics, Rubbers and the Environment

Plastic and rubber feedstocks (polymers) use only 4-5% of oil and gas, and processing of polymers into plastic materials needs a maximum of an another 2-3% oil, so in total, 6-8% of oil and natural gas is being consumed for their production. It is worth stressing that 'most of the oil and natural gas is being used for energy and transportation', but not for the production of plastics and rubbers. Worldwide reserves of petroleum are currently estimated as 200 billion tons, which is enough for about 50 years at the existing rate of consumption, and the natural gas supply is also estimated to be sufficient for the same duration of time. Although the share of plastics and rubbers in this consumption are rather low, as explained previously, after a certain time limit (50 years?) the natural resources will not be enough for their production. Hence, the immediate task will be the exploration of different convenient ways for their production for the near future, which will rest on other renewable resources (other than oil and gas), with the most proper 'eco-profiles' and with the smallest impact on the environment (green plastics and rubbers).

One should remember that, energy profiles for the production of polymers are known to be one of the lowest and are the most favourable (3.1 kWh/kg), they are much less labour and energy intensive as compared to other traditional materials known (74 for aluminium, 8 for glass, 7 for paper), hence, their share of the use of oil and gas as energy during the processing of plastics and rubber are also comparably small. In addition, the manufacture of one of the most traditional materials (i.e., paper) is more polluting environmentally, than the manufacture of polymers [11].

9.3.2 Plastics and Rubbers Waste

The overall benefits of plastics and rubbers have been questioned for a long time because of their ultimate disposal problems. The amount of plastics in municipal solid waste is about 18% by volume, and they last for a long time in the environment. If all other organic wastes - such as paper, food waste and wood are excluded, plastics constitute about 30% of the remaining weight, or half the remaining volume. It is estimated that, over 30 million tons of plastics are discarded yearly into waste in the US, and 20 million tons in the EU. Almost one-third of this waste comes from packaging (plastics which are discarded after use with a useful lifetime of less than one year), plus there are those from plastic construction materials (with useful life times much longer than a year), and others. Which means that, on the average, about 50 million tons of plastic material ends up in waste within two years of their production, worldwide.

Plastic litter is rather long lasting and can stay intact for a very long time without any deterioration or degradation. There are techniques already being developed and used to manage plastics waste by their proper recycling. Recycling constitutes the final stage in the common triangle of 'Reduce-Reuse-Recycle'. The 'reduction', meaning the 'source reduction', in this triangle, includes several proposals for replacement of plastics (i.e., packaging materials) with others (paper, glass), which is not a very attractive approach, if the gains and losses are considered more closely (i.e., glass consumes too much energy for its processing and hence more gas and oil needs to be used, while a shift to paper use means that more wood and more trees are needed (neither are sustainable at all). Detailed 'impact analysis' to compare a hot drink paper cup and a polystyrene (PS) foam plastic cup showed that, use of the first is not more favourable than the second, if their environmental impacts (their resource and energy use, as well as littering) are considered [12, 13].

The 'reuse' of once used plastics, may not be always possible and has certain limitations, because of impurities and health hazards of the chemicals involved. Hence 'proper effective recycling', although it still has a number of limitations, remains a promising approach. This option is still under development. One of the most difficult issues in plastics recycling still remains the recycling of composite plastics and laminates.

In recycling, there are three main methods employed: 'physical recycling, chemical recycling and incineration'.

In physical recycling, used plastics are reprocessed directly back to produce new (generally inferior quality) materials, for less demanding applications. The waste material already had certain additives from its previous history, and during its re-processing, new additives are added to the system. Hence, the new, reprocessed material is expected to contain known and unknown hazardous substances, and it should not be used for food or medical applications at all. In chemical recycling, the plastic is transformed into other chemicals that can be used as monomers or intermediates. In incineration (pyrolysis) also known as energy recycling, the energy of plastic is used by its careful incineration in special ovens.

9.3.2.1 Biodegradable Polymers, UV Degradable Polymers

'Waste minimisation' in the plastics and rubber sector is one of the main issues in achieving the sustainability, as discussed previously, and use of biodegradable, or simply degradable polymers, will be one of the most meaningful ways to do this. Polymers are of two types, as regards to their degradabilities. They are either: (a) already inherently degradable (bio- or UV degradable), or (b) non-degradable. For the latter case, however, they can still be made easily degradable by the use of proper additives or reactions. In the following part, these two methods will be outlined briefly. Once the polymer gains this property (ease of degradation in the environment), it can degrade with the help of certain micro-organisms and fungi that exist in soil (or by sunlight), when the polymer is left alone as waste in the environment. By definition, a polymer is usually considered biodegradable if all its mass breaks down in soil or water into carbon dioxide and water within six months, and in principle, all polymers that can be oxidised or hydrolysed should be biodegradable ultimately [14].

a) As far as biodegradability of plastics is concerned, it is known that some synthetics and natural polymers are **inherently biodegradable** already (bioplastics). Examples of biodegradable synthetics are: polyvinyl alcohol (PVA), polycaprolactone, polyglycolic acid, polyethylene oxide and polybutylene succinate ionomers, while polylactic acid (PLA) is a natural biodegradable material. Biodegradable PLA/chitosan composites are also prepared to balance the early easy degradability of PLA [15]. When these are put in the waste stream, they begin to self-destruct with the help of microorganisms by biotic chemical degradation, and the degraded parts return back to the eco-system, without harming much of the environment.

There are also degradable synthetic polymers which are obtained by synthesis: in these, there are mostly UV sensitive groups bonded covalently to the main polymer chain. Companies like DuPont, Union Carbide and Dow, are producing certain special ethylene-carbon monoxide copolymers of polyethylene (PE) with carbon monoxide. While for the so called Ecolyte or EcoPlastics, a number of different polymers (such as PE, polypropylene and PS) are prepared by incorporating ketone molecules into the polymer chain, all of which have been shown to undergo photodegradation in 60-600 days of outdoor exposure easily and spontaneously [16-19]. In an another study, it has been shown that a completely environmentally degradable plastic can be made by copolymerising epoxide and carbon dioxide, catalysed by a rare earth

191

compound [14], or the biodegradable elastomer, poly(1,8-octanediol-*co*-citric acid). This citric acid based biodegradable elastomer are used for medical applications, it is highly biocompatible and is specifically designed for tissue engineering [20].

Using special reactions, it has also been shown that, a degradable polymer can be produced as well i.e., by transesterification, new biodegradable polyesters can be synthesised from polybutylene adipate-*co*-succinate and polyethylene terephthalate [21]).

There are also biodegradable composites based on proteins and homogeneous, transparent, strong, water resistant, highly permeable to water vapour and highly gas selective biodegradable materials mainly used for food, and medical applications, in addition to various packaging, agriculture, and controlled release systems, have been obtained [22]. However, there are no completely harmless sustainable thermosets available yet, either for partially biological origin as well as for completely biological origin thermosets (i.e., if a phenolic resin from the shell of cashew nnuts (cashew nut shell liquid), is reacted with formaldehyde of wood alcohol origin).

b) **Common synthetic polymers are inherently non-biodegradable,** and they *are claimed be made biodegradable* by use of certain additives that attract micro-organisms, such as starch, (starch-based polymers). Starch is added to PE formulations (sold under the name 'Ecostar', with 6-80% starch loadings), as well as to saturated polyesters to introduce biodegradation, and with PVA, and polycaprolactone, to increase their existing inherent biodegradations for the latter. These starch-based polymers have initiated an on-going international discussion about their effectiveness (i.e., plastic bags made from these blends are claimed not to degrade in landfills properly, degrading the starch and leaving the polymer undegraded). There are studies done with other additives as well, such as, use of biodegradable soy proteins (used with saturated polyesters, where glycerol is used as a compatibilising agent). The new biodegradable starch-based polyesters have excellent properties at a low cost [23].

Already there are certain restrictions for certain non-biodegradable plastic packaging materials (the US Plastics Pollution Research and Control Act of 1987, Public Law 100-220 and the Annex of the MARPOL (marine pollution) Convention – the International Convention for the Prevention of Pollution from Ships which prohibit the disposal of plastics at sea, which allowed the US Navy promote the development of aquatic biodegradable plastics at sea.

9.3.3 Polymers from Natural Renewable Sources (Sustainability Through Green Polymers)

The second important basic step in achieving sustainability for plastics and rubbers will be their production and synthesis from renewable sources, namely they should be transformed into, so-called, green (smart) plastics and rubbers [22].

The use of renewable resources for producing a new generation of sustainable polymers (*via* green chemistry) and the production of bio- or environmentally-degradable plastics with proper chemical and physical properties, with considerable reduction of health hazards and risks are the ideal scenario to aim for in the near future. In fact, some new generation polymers can be produced from renewable sources and they are (or they can be made) bio or environmentally - degradable. For the time being, there is one main problem that really prevents the growth of renewable polymers, which is the economics: they are rather expensive.

Most of the polymers we are familiar with are derived from non-renewable fossil resources, (i.e., crude oil and natural gas), and they produce non-(bio)degradable solid wastes at the end of their useful life times. Legislation in the EU and other countries are looking for ways of new alternative, more sustainable ways of producing and disposing of polymers. And there is one alternative: the possibility of substituting the non-biodegradable polymers derived from non-renewable resources with renewables (green polymers). Green polymers are made from feedstocks other than fossil feedstocks, they pose less or no waste management problems, and are environmentally friendly, with no intrinsic hazards to human health and the environment. Any renewable feedstock only after evaluation for its chronic toxicity, carcinogenicity, ecotoxicity, can be used for green chemistry. Because if the starting material possesses any hazard to human health and to the environment, this effect most probably will be felt throughout the life cycle of the chemical product.

Carbohydrates (mono, oligo, or polysaccharides, such as cellulose, starch, chitin), oils and fats both of plant and animal origin (triglycerides and oleic acid) and proteins (wool, silk, keratin); are some of the natural biopolymers and are examples of renewable sources that can be (and are being) used in the production of green polymers. During the Winter Olympics in Norway, the thermoplastic starch of potato or cornstarch origin was used successfully for producing disposable plates and knives.

There are also synthetic green biopolymers, which are synthesised either:

(i) In the laboratory from monomers existing in nature, which are also called *synthetic* or *honorary biopolymers*.

In fact, PLA produced from corn starch dextrose or lactic acid, is probably the first commercial scale production of a green polymers [24-27]. PLA is a polyester and a thermoplastic. It is interesting to note that, PLA has similar processability properties as PS, and similar tensile properties as polyethylene terephthalate, with good resistances to fats and oils, and has good a drape. PLA is already used mainly in the packaging and in textile sector. As another example, a number of vegetable oils have already been tested (and are being used) as feedstocks for green polyurethane (PU) polymers [17]. Polyols (feedstocks for PU) are successfully being synthesised from oleochemicals [28]. Soy-based PU formulations are used for carpet-backing mainly (Biobalance), and a new polyester resin from soybeans and corn (Envirez) are already available. Amino acids (such as aspartic acid) can be polymerised to produce polyacrylate-like synthetics, triglycerols, or triglycerides, which are found in plants and animals and can be used to produce drying, semidrying and non drying polymeric oils, by epoxidation.

(ii) They can also be synthesised by microbially (*microbial-synthesised biopolymers*) using certain micro-organisms through bacterial fermentation (usually of sugars). This second technique is known as *biocatalytic polymer formation*, which is another approach for sustainable production of green polymers. Biopol (Monsanto) is an example, which are natural aliphatic polyesters, polyhydroxyalkanoates, as typified by polyhydroxybutyrates and polyhydroxyvalerates), i.e., poly 3-hydroxybuyrate-*co*-3-hydroxyvalerate), and they certainly are renewable [29]. Biopol-based coffee cups, shampoo bottles, sacks and other disposables are available and have been used for some time. A number of similar polymerisations are found in the literature, such as, the polymerisation of lactones to polyesters [30], acrylamide production [31], and production of certain (green) conducting polymers from enzymic polymerisation of phenols and anilines [32], protein-based thermoplastic elastomers [33], are strong candidates as novel scaffolds for tissue engineering and for controlled drug release and cell encapsulations.

Currently green biopolymers are not used extensively, mainly due to economic as well as some technical reasons, however, they have good prospects for the future.

9.3.4 Sustainability Through Additives

Most of the additives used in plastics and rubbers are toxic and during their production these are also considered as non-environmentally friendly. To have sustainable development with sustainable compounds, these additives should be replaced by their more sustainable counterparts.

In the following sections, some of these will be presented briefly.

9.3.4.1 Sustainable Plasticisers

There are plasticisers developed from renewable sources that are degradable. The Hallgreen plasticiser (diisooctyl sebacate; RTD Hallstar Inc) is one such, which is mainly certain ester compounds, and can replace the conventional petroleum-based plasticisers. It does not contain any toxic substances and is completely safe for humans, and also completely biodegradable. Already a number of examples of green, degradable, sustainable plasticisers have been developed and there are also a number of ongoing studies. The most critical one should be the one to replace phthalates. Yearly about

4.5 M tonnes of different phthalates are currently being used worldwide as the plasticiser for polyvinylchloride (PVC), which needs to be replaced by a sustainable counterpart, probably soybean oils. Research teams from the Institute for Agrotechnology and Food Innovation and the Wageningen University (Netherlands) are working on a project to develop a 'green' alternative to phthalates, which is based on sugar alcohols. Meanwhile, polymeric plasticisers, i.e., polyesters based on adipic acid, are successfully used in PVC as a sustainable replacement of phthalates. Their migration is negligible, with very low volatilities. Esters of fatty acids and monocarboxylic acids are potential 'secondary plasticisers' for PVC compounds. Stearic acid esters are green plasticisers and processing agents in general, and are also used as lubricants for PS, with food-contact approval. Sebacates and adipates can be used as plasticisers (for PVC) and they have food-contact approvals, however, di-butyl sebacate is more efficient at lower temperatures. Epoxidised grades (of soya bean oil, linseed oil) can also be used as plasticisers (in PVC, in alkyd resins and chlorinated paraffins), and as pigment dispersing agents in plasticised PVC. Alkyl epoxy stearates as plasticisers provide good low temperature properties [34].

9.3.4.2 Sustainable Lubricants, Colorants and Flame Retardants

Sustainable external lubricants (mould release agents) are mainly waxes, such as bee wax, Carnauba wax and rapeseed oil, where boiling points are critical [35]. As long as the boiling point of the sustainable external lubricant is not lower than the processing temperature of the melt, it can be considered to be a good candidate.

Most of the sustainable colorants are either mineral (such as carbon black, iron oxide, graphite and titanium dioxide), or of vegetable origin. At this time, mostly mineral colorants are being promoted, up to levels of 49% [36].

All flame-retardants have certain health hazards to some degree that changes from substance to substance, with the exception of alumina trihydrate, which is not toxic and functions by limiting the propagation of combustion *via* releasing the water of hydration. However, although this compound is sustainable, it also brings one drawback: they must be applied in rather large proportions, which can interfere with the biodegradabilities [35].

References

1. *Approaches to a Sustainable Development (Global Development and the Environment*, Eds., R.M. Auty and K. Brown, Routledge, London, UK, 1997.

2. M. Yakowitz, *Sustainable Development: OECD Policy Approaches for the 21st Century,* OECD Publications, Paris, France, 1997.

3. A. Azapagic in *Handbook of Green Chemistry and Technology*, Eds., J. Clarke and D. MacQuarrie, Blackwell, Oxford, UK, 2002, p.62.

4. N. Yoda, *Journal of Macromolecular Science, Part A: Pure and Applied Chemistry*, 1996, **33**, 12, 1807.

5. D. McLaren, S. Bullock and N. Yousuf, *Tomorrow's World: Britain's Share in a Sustainable Future*, Earthscan Publications, London, UK, 1998.

6. *Eco-Design of Energy-Using Products*, Enterprise and Industry, The European Commission, http://ec.europa.eu/enterprise/eco_design/index_en.htm

7. *Eurostat*, The European Commission, http://europa.eu.int/comm/eurostat/sustainabledevelopment [Link doesn't work]

8. *Europa*, EU Press Room, The European Union, http://europa.eu.int/comm/press_room/presspacks/health/pp_health_en.htm

9. *ROHS*, Directive http://www.rohs.gov.uk/

10. *EuropaEnvironment*, The European Commission, http://ec.europa.eu/environment/waste/weee_index.htm

11. G. Scott, *Polymers and the Environment*, The Royal Society of Chemistry, Cambridge, UK, 1999.

12. M.B. Hockings, *Science,* 1991, **251**, 4993, 504.

13. E.S. Stevens, *Green Plastics: An Introduction to the New Science of Biodegradable Plastics*, Princeton University Press, Princeton, NJ, USA, 2002.

14. J.P. Gao and H.M. Xiang, *Tianranqi Huagong*, 2004, **29**, 2, 55.

15. N. Saha, M. Zatloukal and P. Saha, *Polymers for Advanced Technologies*, 2003, **14**, 11-12, 854.

16. H. Omichi in *Handbook of Polymer Degradation*, Eds., A. Maadhah, S. Halim Hamid and M.B. Amin, Marcel Dekker Ltd, New York, NY, USA, 1992, p.335.

17. D.K. Platt, *Biodegradable Polymers Market Report,* Rapra Technology, Shawbury, UK, 2006.

18. C. Bastioli, *Handbook of Biodegradable Polymers*, Rapra Technology, Shawbury, UK, 2005.

19. R. Smith, *Biodegradable Polymers for Industrial Applications*, Woodhead Publishing, Cambridge, UK, 2005.

20. J. Yang, A.R. Webb and G.A. Ameer, *Advanced Materials*, 2004, **16**, 6, 511.

21. S.W Kim, L. Jeong-Cheol, K. Dae-Jin and S. Kwan-Ho, *Journal of Applied Polymer Science*, 2004, **92**, 5, 3266.

22. S. Guilbert in *Proceedings of ACS Polymeric Materials: Science and Engineering*, Anaheim, CA, USA, 2004, Volume 91, p.110.

23. D. Graiver, L.H. Waikul, C. Berger and R. Narayan, *Journal of Applied Polymer Science*, 2004, **92**, 5, 3231.

24. H.C. James and J.E. Hardy Jeffry, *Sustainable Development in Practice: Case Studies for Engineers and Scientists*, Eds., A. Azapagic, R. Clift and S. Perdon, Wiley, Chichester, UK, 2004

25. J. Clark and D. Macquarrie, *Handbook of Green Chemistry and Technology*, Blackwell Publishing, Oxford, UK, 2002.

26. *ENDS Report*, 2000, No.300, 19.

27. A.K. Kulshreshtha, *Popular Plastics & Packaging*, 2004, **49**, 6, 91.

28. L. White, *Urethanes Technology*, 2006, **23**, 2, 34.

29. L.A. Madden, A.J. Anderson and J. Asrar, *Journal of Macromolecules*, 1998, **31**, 17, 5660.

30. S. Kobayashi, *ACS Polymeric Materials Science and Engineering*, 1996, **74**, 32.

31. H.L. Holland in *Handbook of Green Chemistry and Technology*, Eds., J. Clarke and D. Macquarrie, Blackwell Publishing, Oxford, UK, 2002, p.188.

32. L.A. Samuelson, A. Anagnonstopoulos, K.S. Alva, J. Kumar and S.K. Tripathy, *Macromolecules*, 1998, **31**, 13, 4376.

33. K. Nagapudi, W.T. Brinkman, J. Leisen, B.S. Thomas, E.R. Wright, C.Haller, X. Wu, R.P. Apkarian, V.P. Conticello and E.L. Chaikof, *Macromolecules*, 2005, **38**, 2, 345.

34. P.S. Duane, *Omnexus Trend Reports*, 2006.

35. N. Tucker and M. Johnson, *Low Environmental Impact Polymers*, Rapra Technology, Shawbury, UK, 2004.

36. M.J.A. Van Den Oever, C.G. Boeriu, R. Blaauw and J. Van Haveren, *Journal of Applied Polymer Science*, 2004, **92**, 5, 2961.

Some Additional Related References

1. G. Woodward, G.P. Otter, K.P. Davis and K. Huan, inventors; Rhodia Consumer Specialties Limited, assignee; WO 4056886, 2004.

2. M. Hakkarainen and A.C. Albertsson, *Advances in Polymer Science*, Springer, Heidelberg, Germany, 2004.

3. L. Zan, L. Tian, Z. Liu and Z. Peng, *Applied Catalysis A: General*, 2004, **264**, 2, 237.

4. L. Bosiers and S. Engelmann, *Kunststoffe*, 2003, **93**, 12, 48.

5. S.Fisher, E. De Craenmehr, J.J. De Vlieger and T.M. Slaghek, inventors; Nederlandse Organisatie voor Toegepast Natuurwetenschappelijk Onderzoek TNO, assignee; WO 4029147, 2004.

6. J. Pal, H. Singh and A.K. Ghosh, *Journal of Applied Polymer Science*, 2004, **92**, 1, 102.

7. J.H. Kim, *Journal of Industrial and Engineering Chemistry*, 2004, **10**, 2, 278.

8. E.S. Stevens, *Green Plastics*, Princeton University Press, Princeton, NJ, USA, 2002.

9. C. Billie, *Green Composites*, Woodhead Publishing, Cambridge, 2004.

10. J.T. Anastas and J.C. Warner, *Green Chemistry: Theory and Practice*, Oxford University Press, Oxford, UK, 2000.

11. H. Stapert in *Polymer Products and Waste Management: A Multidisciplinary Approach*, Ed., M. Smits, International Books, Boston, MA, USA, 1996.

10 List of Some Health Hazard Causing Solvents, Monomers and Chemicals Common for Plastics and Rubbers

Plastics and rubbers are not used alone in their pure states, but usually they contain various chemicals which may have potential health hazards for humans. These chemicals can be either those that exist initially in the system and may be left after polymerisation (such as, monomers, catalysts), or those that are added afterwards during processing (i.e., plasticisers, stabilisers).

This chapter will give brief information on some of the chemicals which are most commonly used for plastics and rubbers.

ACETALDEHYDE (CH_3CHO, SMILES CC=O)

Alternative Names/Abbreviations: Ethanal, acetic aldehyde, ethyl aldehyde, methyl formaldehyde

CAS Registry Number: 75-07-0

Details: A colourless liquid, used in the production of polyester resins.

Toxicology: Acetaldehyde is a suspected human carcinogen (A3 animal carcinogen), and genotoxic substance, and an irritant to eye, skin and respiratory tract. Exposure to 130 ppm for 30 minutes, causes mild upper respiratory irritation. Exposure to 50 ppm for 15 minutes, causes mild eye irritation [1]. Exposure to 25 ppm for 15 minutes, causes eye irritation (for sensitive objects) [2].

Ceiling threshold limit value (TLV-C) by the American Conferences of Governmental Industrial Hygienists (ACGIH), 2003: 25 ppm (equivalent to 45 mg/m³).

ACETAMIDE (CH_3CONH_2, SMILES CC(=O)N)

Alternative Names/Abbreviations: Acetic acid amide, ethanamide

CAS Registry Number: 60-35-5

Details: Solid (crystals), a plasticiser.

Toxicology: Acetamide is an irritant (to mucous membranes), a toxin (for liver), and a suspected human carcinogen (a proved animal carcinogen).

No TLV-C value is established [3].

ACETIC ANHYDRIDE [(CH$_3$CO)$_2$O, SMILES CC(=O)OC(=O)C]

Alternative Names/Abbreviations: Acetic oxide, acetyl oxide, ethanoic anhydride, acetic acid anhydride

CAS Registry Number: 108-24-7

Details: A colourless liquid, used in the production of plastics.

Toxicology: Its vapour is an irritant above 5 ppm to eyes and mucous membrane, and it is a strong corrosive (if it is liquid, to skin).

Threshold Limit Value/Time Weighted Average value (TLV-TWA) by ACGIH, 2003: 5 ppm (equivalent to 21 mg/m^3) [4].

ACETONITRILE (CH$_3$CN, SMILES CC#N)

Alternative Names/Abbreviations: Methyl cyanide, cyanomethane, ethanenitrile, AN

CAS Registry Number: 75-05-8

Details: A colourless, volatile liquid, a chemical intermediate in the production of plastics.

Toxicology: At low concentrations, acetonitrile causes headache and nausea, whilst at high concentrations, causes convulsions and death. Acetonitrile is toxic and flammable. It is metabolised into highly toxic HCN and thiocyanate. The main toxic effects of acetonitrile are attributed to the metabolic release of cyanide, in fact, specific cyanide antidotes are used in acetonitrile poisonings [5-7].

TLV-TWA value by ACGIH, 2003: 40 ppm (equivalent to 67 mg/m^3) [6, 7].

ACROLEIN (C_3H_4O, SMILES C(=O)C=C)

Alternative Names/Abbreviations: Acrylaldehyde, 2-propenal, allyl aldehyde, propylene aldehyde, aqualin

CAS Registry Number: 107-02-8

Details: A colourless liquid, used in the production of acrylic acid and its resins.

Toxicology: Acrolein is a strong irritant for the upper airways (with an irritation threshold of 0.25-0.5 ppm), eyes, as well as for skin. Skin exposure causes serious damage. It has been used as a chemical weapon during WW1, however, it is still not outlawed by the Chemical Weapons Convention. Acrolein concentrations of 2 ppm are immediately dangerous to life; 150 ppm is lethal after 10 minutes [8]. Acrolein is a suspected human carcinogen. Connections have been found between acrolein (found in tobacco cigarettes and certain cooking oils as well as the chemical itself) and the risk of lung cancer.

TLV-TWA value by ACGIH, 2003: 0.1 ppm (equivalent to 0.23 mg/m³).

ACRYLATE POLYMERS

They are a group of polymers also commonly known as **acrylics** or **polyacrylates**. They have high transparencies and resistance to breakage, when compared to conventional (window) glass.

Some acrylate monomers are acrylic acid, butyl acrylate, 2-ethylhexyl acrylate, methyl acrylate, ethyl acrylate, acrylonitrile, *n*-butanol, methyl methacrylate (plexiglas) and trimethylol propane triacrylate (TMPTA).

Superglue is a formulation of cyanoacrylate.

ACRYLAMIDE (C_3H_5NO, SMILES C=CC(N)=O)

Alternative Names/Abbreviations: AA, acrylic amide, propenamide, ethylene carboxamide, vinyl amide

CAS Registry Number: 79-06-1

Details: Solid (powder), monomer of polyacrylamide, and a crosslinker (for vinylic polymers). Acrylamide may exist in fried or baked foods (mainly as a result of browning or overcooking, because none of the uncooked food contains it, and especially in baked and fried starchy foods, such as potato chips and bread), as well as in olives and prune juice. A lawsuit was filed in the US in August 2005 by California Attorney General, Bill Lockyer against top french fries and potato chips makers, asking them to warn consumers of the potential risks involved [7]. Smoking also produces acrylamide. Acrylamides can also be created during microwaving.

Toxicology: Acrylamide is an irritant (on skin), a strong poison affecting central and autonomic nervous systems and the male reproductive system (genotoxic); and an A3 suspected human carcinogen (a proved animal carcinogen).

Worksafe Australia recommends an 8 hour time weighted average (TWA) exposure limit of 0.03 mg/m^3.

TLV-TWA value by ACGIH, 2003: 0.03 ppm (with a notation for skin absorption) [6, 7].

ACRYLIC ACID [C$_3$H$_4$O$_2$ SMILES C=CC(=O)O]

Alternative Names/Abbreviations: 2-Propenoic acid, acroleic acid, ethylene carboxylic acid, vinylformic acid

CAS Registry Number: 79-10-7

Details: A colourless liquid, a starting material for making acrylates (with amide and acrylonitrile monomers) and other polymers and copolymers (with vinyl, styrene and butadiene), medical and dental materials.

Toxicology: Acrylic acid is a strong irritant for eyes, skin and upper airways, with no indication of its systemic toxicity or carcinogenicity. Inhalation may be fatal because of spasm, inflammation and swelling of the bronchi, chemical pneumonitis and pulmonary oedema (fluid in the lungs). It may cause sensitisation (and allergic reactions). It is metabolised rapidly to carbon dioxide in the human body.

TLV-TWA value by ACGIH, 2003: 2 ppm (equivalent to 5.9 mg/m^3), with a notation for skin absorption.

ACRYLONITRILE (C$_3$H$_3$N, SMILES C=CC#N)

Alternative Names/Abbreviations: Cyanoethylene, propenenitrile, vinyl cyanide, ACN

CAS Registry Number: 107-13-1

Details: A colourless liquid, monomer for polyacrylonitrile and other synthetics such as styrene acrylonitrile *co*-polymer.

Toxicology: Acrylonitrile is an irritant for the eyes, skin (may lead to second degree burns) and upper airways, with somewhat non-specified systemic toxicity (maybe to the central nervous, cardiovascular and hepatic systems). Acute exposure first leads to weakness, then to asphyxia and to death [9]. *In vitro* genotoxic studies showed gene mutations and related abnormalities. ACN is highly flammable and toxic, undergoing explosive polymerisation. The burning of ACN releases fumes of highly toxic HCN and oxides of nitrogen. Acrylonitrile is classified as a possible human carcinogen (IARC group 2B), and a proved animal carcinogen. ACN is known to metabolise to cyanide in the human system.

TLV-TWA value by ACGIH, 2003: 2 ppm (equivalent to 4.3 mg/m^3), with a notation for skin absorption.

ALLYL CHLORIDE, (C$_3$H$_5$Cl, SMILES C=CCCl)

Alternative Names/Abbreviations: Chlorallylene, 3-chloroprene, 3-chloropropene, 1-chloro-2-propene, 3-chloro-propylene, 2-propenyl chloride

CAS Registry Number: 107-05-1

Details: A liquid, starting material for making epichlorohydrin and epoxy resins.

Toxicology: Allyl chloride is toxic and flammable. It is an irritant for the eyes (between 50-100 ppm), skin and upper airways (below 25 ppm). Its chronic exposure may cause toxic polyneuropathy. Allyl chloride is detectable below 3 ppm [10].

TLV-TWA value by ACGIH, 2003: 1 ppm (equivalent to 3 mg/m^3).

ALLYL GLYCIDYL ETHER (C$_6$H$_{10}$O$_2$, SMILES C=CCOCC1CO1)

Alternative Names/Abbreviations: Allyl 2, 3-epoxypropyl ether, AGE

CAS Registry Number: 106-92-3

Details: a liquid, used as a starting material for rubber and a reactive diluent for epoxy resins.

Toxicology: Allyl glycidyl ether is an irritant for eyes, skin (contact dermatitis) and upper airways, with some evidence for it being an animal carcinogen.

TLV-TWA value by ACGIH, 2003: 1 ppm (equivalent to 4.7 mg/m^3).

ALUMINUM

Alternative Names/Abbreviations: Alumina, Al

CAS Registry Number: 7429-90-5

Aluminum is a non essential element. Aluminum in the body is difficult to remove, but it is not regarded as a poison at normal levels (approximately 60 mg in the whole body).

Aluminum helps to stop bleeding, but can cause brain damage at high levels. Aluminum was believed to be the main cause of Alzheimer's disease, however, it has recently been shown that this may not be true. Metallic aluminum is a good conductor of electricity (and is used in the electrical cable industry), and a good reflector (used largely in insulation applications, for heat reflecting blankets, in solar mirrors and in plastic packaging as a coating).

Common Related Aluminium Compounds: Aluminum occurs in nature as alum (potassium aluminum sulfate) which is used largely as a fixing agent for dyeing, and by doctors to stop bleeding, as well as a preservative in paper industry.

Aluminum oxide is a flame retarder and hence is a flame retarding additive for plastics.

Toxicology: TLV-TWA value of aluminum is by ACGIH, 2003: 10 mg/m^3 for the metal dust as aluminium, and 2 mg/m^3 for the soluble salts and alkyls as aluminium.

4-AMINODIPHENYL (C$_{12}$H$_{11}$N, SMILES NC1(=CC=C(C=C1)C2=CC=CC=C2))

Alternative Names/Abbreviations: Para-aminodiphenyl, 4-aminobiphenyl, biphenylamine, para-xenylamine), 4-ADP

CAS Registry Number: 92-67-1

Details: Solid (crystals), a previous rubber antioxidant (no longer in commercial production).

Toxicology: 4-Aminodiphenyl enters the body mostly through skin contact or inhalation, and it is determined by IARC and ACGIH as an A1 human carcinogen (causing bladder cancer), and it is regarded as the most hazardous within aromatic amines for its carcinogenic potential to humans [11], and it is strongly suggested that contact by any means should be avoided.

There is no assigned ACGIH threshold limit value available for 4-aminodiphenyl [12].

AMMONIUM CHLORIDE (FUME) (NH$_4$Cl, SMILES [H][N$^+$]([H])([CH])[H].[Cl$^-$])

Alternative Names/Abbreviations: Ammonium muriate fume

CAS Registry Number: 12125-02-9

Details: An odourless fume, used as a hardener for formaldehyde-based resins.

Toxicology: A mild (to severe) irritant to the eyes, and to the upper airways when inhaled. No data on its toxic effects are available.

TLV-TWA value of ammonium chloride fume is by ACGIH, 2003: 10 mg/m^3).

n-AMYL ACETATE (CH$_3$COOC$_5$H$_{11}$, SMILES CCC CCOC(=O)C)

Alternative Names/Abbreviations: Amyl acetic ether, pentyl acetate

CAS Registry Number: 628-63-7

Details: a liquid, used as solvent for lacquers, paints, leather and so on.

Toxicology: *n*-Amyl acetate is an irritant for mucous membranes (Exposure of 200 ppm for 3-5 minutes produces eye/nose and throat irritation).

TLV-TWA value of *n*-amyl acetate is by ACGIH, 2003: 100 ppm (532 mg/m^3).

ANTIMONY (Sb) AND ANTIMONY COMPOUNDS

CAS Registry Number: 7440-36-0

Antimony is a non essential element in the human body and it has no biological role. Its total amount existing in the body normally is about 2 mg, and 100 mg is the lethal limit. Antimony is very toxic, but it has also been prescribed as a treatment for different kinds of diseases for hundreds of years. For example antimony has been used to treat several parasitic diseases such as Schistosomiasis caused by several species of flatworm in both humans and animals. Antimony is thought to attach itself easily to sulfur atoms of certain enzymes which are used by both the parasites and human body, and in small doses can kill the parasite without causing any damage to the patient.

In fact, *in small doses*, it stimulates the metabolism. *At higher doses*, it causes vomiting and damages the liver.

Once in the body, antimony cannot be easily removed. Its preferential permanent attraction is to the sulfur atoms (which are the enzyme's active sites and is probably this the cause of its main poisonous action).

Metallic antimony was used as a re-usable laxative (to be taken orally in the form of small tablets) for centuries.

Common Related Antimony Compounds: Within the antimony compounds, there are *antimony hydride* (SbH_3, stibine, a gas) which is the most deadly compound, followed by *antimony sulfide* (Sb_2S_3, stibi), which is used as a medicament for skin complains and burns. *Antimony sulfites* are used in rubber compounding, and in camouflage paints because of its reflection of infrared. *An antimony salt of tartaric acid*, known as *tartar emetic*, is prepared by leaving some wine in a cup made of antimony overnight (this compound was then used by vets and doctors, mainly to expel bad humours from the body).

About 5 mg of antimony shows a strong diaphoretic action (induces sweating), while 50 mg or more acts as an emetic. There are rumours that Mozart's death was due to acute antimony tartarate poisoning. It is also known that several Victorian doctors used antimony to dispose quietly of their unwanted wives or relatives. *Antimony trioxide* is used as a flame retardant in plastics, while *antimony chloride* is used as a catalyst and a colouring agent.

Antimony oxide/lead oxide - carbonate mixture is a well known pigment and paint (Naples yellow), and recently it has also been used as a flame retardant for plastics [primarily in polyvinyl chloride (PVC)] for their specific use in car components, in TV casings and so on, because it quenches the fire by reacting chemically with burning materials.

Antimony/lead alloys are used in bearings and electrical cable sheathing.

Toxicology: *Antimony trioxide* causes pulmonary injury and is considered as carcinogenic for animals and an A2 suspected human carcinogen, and *antimony hydride* is highly toxic. The use of *antimony oxide/lead oxide and carbonate* mixture in PVC was criticised in the 1990s for being the cause of *'cot mattress deaths'* of babies. It was believed that antimony in the mattress was converted to the volatile deadly poison *stibine* by a particular fungus, existing usually in mattresses. This claim was discounted by a special panel in 1998.

However, it was found that still there may be high levels of antimony in some old houses, probably due to antimony containing lead piping systems and layers of antimony containing lead paints.

Antimony trisulfites are cardiotoxic.

TLV-TWA value for antimony and its compounds is by ACGIH, 2003: 0.5 mg/m^3 as antimony.

206

ARSENIC (As) AND ARSENIC COMPOUNDS

CAS Registry Number: 7440-38-2

Metallic arsenic is a grey brittle metal, and it is known as one of the most deadly poisons, and it is *one of the few compounds (besides vinyl chloride) that causes the rare liver cancer angiosarcoma,* but it is still thought of as an essential (trace) element for some animals and for humans, with a necessary intake of 0.01 mg per day, most probably due to the special metabolism of certain amino acids (chickens and rats fed with an arsenic free diet are found to have their growth inhibited). It is claimed that arsenic in small doses stimulates the metabolism and boosts the formation of red blood cells. In fact, its derivatives can be used illegally as a dopant for racehorses and even to fatten poultry and pigs.

In the past, beginning from the age of Hippocrates, arsenic compounds (mainly arsenic sulfide) were prescribed by doctors as a 'medicine' to cure a series of ailments (rheumatism, malaria, tuberculosis, diabetes, syphilis).

In the nineteenth century, another compound of arsenic, *potassium arsenite*, was considered to be a general tonic and as an aphrodisiac and it was customary to use it by adding a few drops of it to a glass of water or wine.

Common Arsenic Compounds: Arsenic trichloride (a liquid), arsenic trioxide (a crystalline solid) are the two common compounds of arsenic, the latter being used as wood preservative as well as in paints and wall papers.

Toxicology: Arsenic compounds are irritants for the skin, upper airways and eyes, leading to gastrointestinal and vascular diseases and various cancers. Inorganic arsenicals and trivalent forms (arsenic trioxide) are more toxic than the organic and pentavalent forms. Arsenic existing in the human body is normally between 0.5-15 mg, and it's lethal dose is normally 150 mg. The daily intake of arsenic through normal diet, where it is in the form of arseno-betaine, is about 1 mg. The WHO's recommended maximum level of arsenic in drinking water is 0.01 ppm or 0.01 mg/l, which is usually absorbed by tissues and can also be excreted rapidly through urination.

Modern Chinese medicine recognises certain arsenic compounds, and the US-FDA approved the use of arsenic trioxide (as Trisenox) for the treatment of leukaemia.

The toxic effect of arsenic intake at high doses is mainly due to its blocking the function of sulfur containing enzymes by binding to them, with the characteristic symptoms of vomiting, colic, diarrhoea and dehydration. In addition, arsenic is chemically very similar to phosphorus, and partly substitutes for it in biochemical reactions.

Prolonged exposure to fumes of arsenic compounds, mainly *arsenic trioxide can* cause skin and/or lung cancer. Arsenic has been called the '*Poison of Kings*' and the '*King of*

Poisons'. Arsenic and its compounds are especially potent poisons, killing by massively disrupting the digestive system, leading to death from shock.

Although it is usually difficult to diagnose arsenic poisoning, once diagnosed, there is a good chance of survival by using its antidotes. It is interesting to note that regular, periodic intake of arsenic can allow the human body to tolerate large doses in time, and this is probably the reason why some people in some parts of the world are immune to arsenic and show no signs of arsenic related discomfort at all, although their water supply contains dangerous levels of arsenic (i.e., even 600 ppm or 600 mg/l). There is a massive epidemic of arsenic poisoning in Bangladesh, where it is estimated that above 50 million people are drinking groundwater with arsenic concentrations elevated well above the WHO's standard of 50 ppb. Many other countries in South-East Asia, such as Vietnam, Cambodia, and Tibet, also have high arsenic groundwaters.

Arsenic derivatives were used as pesticides, herbicides, insecticides, as chemical weapons (Lewisite), and as a paint (the bright yellow pigment, or royal yellow, which was favoured by the seventeenth century Dutch painters, is known to oxidise slowly to the deadly compound *arsenic oxide* over time), and also in wallpapers (during the nineteenth century, most wallpapers were printed with emerald green containing *copper arsenate*, which can be easily transformed into the deadly *methylarsine* gas, with the help of a special mould developed when walls became damp, causing arsenic poisoning, which is believed what happened to Napoleon when he was imprisoned on the damp island of St Helena), and are still being used mainly in glass making, as a wood preservative and in the semiconductor industry (as *gallium arsenide)*.

Wood has been treated with copper arsenate (CCA timber or Tanalith), which is still in widespread use in many countries.

Arsenic contaminated soil can be effectively cleaned by planting the Chinese ladder fern *(Pteris vittata)* that absorbs the arsenic while growing quickly.

Elemental arsenic and arsenic compounds are classified as toxic (and dangerous for the environment) in the EU under directive 67/548/EEC.

The IARC recognises arsenic and arsenic compounds as group A1 human carcinogens, and the EU lists arsenic trioxide, arsenic pentoxide and arsenate salts as category 1 carcinogens.

TLV-TWA value for arsenic and its inorganic compounds (except arsine) is by ACGIH, 2003: 0.01 mg/m^3 as arsenic.

The arsenic allowed under the EPA's proposed drinking water standard is a maximum of 10 ppb.

Some Additional References for Arsenic

1. *Arsenic and Old Lies*, Chemical Industry Archives,
    http://www.chemicalindustryarchives.org/dirtysecrets/arsenic/1.asp

2. A. Meharg, *Venomous Earth: How Arsenic Caused The World's Worst Mass Poisoning*, Palgrave Macmillan, Basingstoke, UK, 2005.

BARIUM (Ba) AND COMPOUNDS OF BARIUM

CAS Registry Number: 7440-41-7

Barium is a 'non essential' element for humans, and has no biological role. It is an abundant element and humans take about 1 mg per day through foods (carrots, onions, lettuce and oranges). Barium is similar to calcium and is absorbed easily, but it does not have any known useful purpose and function in the body.

The total amount of barium existing in the body is about 22 mg.

Barium can cause an increase in heart beat in humans (ventricular fibrillation) and its soluble salts (like barium carbonate or rat poison) are highly toxic to humans because they can dissolve easily in the stomach acid. However, barium sulfate is not soluble in stomach, hence it is considered to be non-toxic. In fact, it is used for X-ray scans of patients with gastric and intestinal disorders. Barium is mainly used in glass making, in medical diagnostics, in textiles and in oil and gas exploration.

BENZENE (C_6H_6)

Alternative Names/Abbreviations: Benzol, cyclohexatriene

CAS Registry Number: 71-43-2

Details: A colourless liquid, used as a solvent and a starting chemical for polystyrene production.

Toxicology: Benzene is a carcinogen. Its levels are regulated internationally in drinking water. The following limits are examples of benzene limits in drinking water set by different authorities: Canada: 5 ppb (5 µg/kg), EU: 1 ppb, WHO: 10 ppb, US: 5 ppb (in California, New Jersey, and Florida: 1 ppb). The EPA and California set public health goals for benzene of 0 ppb and 0.15 ppb, respectively. Human exposure to high concentrations can be fatal in 5-10 minutes. Acute and chronic exposure affects the central nervous system, and bone marrow leading to anaemia and leukaemia, respectively. Benzene is an A1 confirmed human carcinogen.

TLV-TWA value for benzene is by ACGIH, 2003: 0.5 ppm (corresponding to 1.6 mg/m^3) and TLV-STEL of 2.5 ppm (8 mg/m^3).

BENZIDENE ($C_{12}H_{12}N_2$)

Alternative Names/Abbreviations: 4,4´-Biphenylenediamine, 4,4´-diaminobiphenyl, 4,4´-bianiline

CAS Registry Number: 92-87-5

Details: Colourless, crystalline compound used as a hardener for rubber.

Toxicology: A strongly suspected bladder cancer causing agent (ACGIH classified benzidine as an A1confirmed human carcinogen, and the IARC declared that there is enough evidence for its carcinogenicity to humans). Its entry to the body is either through inhalation or skin absorption.

In a study from China, a 25-fold increase in bladder cancer was reported in 1972 through benzidine exposed workers [13].

2,3-BENZOFURAN [C_8H_6O]

Alternative Names/Abbreviations: Benzofuran, benzo(b)furan, cumarone, cumarone, 1-oxindene

CAS Registry Number: 271-89-6

Details: A liquid, and an intermediate in the polymerisation of coumarone-indene resins used as paints and varnishes and adhesives.

Toxicology: It is believed that humans with kidney or liver diseases are susceptible to the toxic effects of this compound.

No TLV-TWA value is assigned yet.

BENZYL BUTYL PHTHALATE ($C_{19}H_{20}O_4$)

Alternative Names/Abbreviations: BBzP, n-butyl benzyl phthalate, benzyl butyl phthalate

CAS Registry Number: 85-68-7

Details: Benzyl butyl phthalate is mostly used as a plasticiser for PVC (artificial leather, floor tiles and vinyl foams).

BBzP was classified as toxic by the European Chemical Bureau a decade ago and its use has declined rapidly since then.

BENZOYL PEROXIDE [$C_6H_5CO)_2O_2$ or $C_{14}H_{10}O_4$]

Alternative Names/Abbreviations: Benzoyl superoxide, dibenzoyl peroxide, lucidol, oxylite

CAS Registry Number: 94-36-0

Details: Benzoyl peroxide is a granular white solid. It is used as catalyst, and as additive in self-curing polymers.

Toxicology: Exposure is usually by inhalation. Benzoyl peroxide breaks down in contact with skin, producing benzoic acid and oxygen, neither of which are significantly toxic. Benzoyl peroxide is a strong irritant to mucous membranes (above 12.2 mg/m³), and can initiate both primary irritation and sensitisation dermatitis. It is a suspected cancer promoter on skin. IARC concluded that there is no solid evidence for its carcinogenicity.

TLV-TWA value for benzoyl peroxide is by ACGIH, 2003: 5 mg/m³.

BENZYL CHLORIDE ($C_6H_5CH_2Cl$)

Alternative Names/Abbreviations: α-Chlorotoluene, ω-chlorotolouene, chloromethyl benzene

CAS Registry Number: 100-44-7

Details: A liquid, used in the production of cosmetics, dyes and resins.

Toxicology: Exposure is usually by inhalation, and it is a strong irritant to eyes, skin and mucous membranes. At 16 ppm, it becomes intolerable within 1 minute. It has been used as a gaseous warfare agent.

Benzyl chloride can cause genetic mutations and can damage chromosomes.

TLV-TWA value for benzyl chloride is by ACGIH, 2003: 1 ppm (corresponding to 5.2 mg/m³).

BERYLLIUM

Alternative Names/Abbreviations: Glucinium, Be

CAS Registry Number: 7440-41-7

Details: Elemental beryllium is a metallic substance. It is a non-essential element for humans, and has no known biological function in the body, but it is an another very dangerous metal. In the human body, there is about 35 mg of beryllium, most probably obtained through the food chain, which does not affect health.

Toxicology: Beryllium mimics magnesium, an essential element of human nutrition, and its displacement from some key enzymes by beryllium are suspected to be the cause of its poisonous action.

Exposure is usually by inhalation. It's compounds can cause dermatitis, acute pneumonitis, acute lung disease and a specific disease (beryllium disease), that erodes the lungs leading to lung cancer, making it hard to walk, causing severe pain and exhaustion, and usually resulting in a slow, painful death by suffocation.

IARC classifies beryllium and its compounds as a proven A2 carcinogen [14].

TLV-TWA value for beryllium and compounds is by ACGIH, 2003: 0.002 mg/m^3.

BISMUTH (Bi)

CAS Registry Number: 7440-69-9

There is approximately 16 and 200 ppb of bismuth in the blood and bones of a human body, respectively.

Bismuth has no biological role in human body, and poses no environmental threat to living species at normal concentrations. Each day, about 20 mg of bismuth is taken into the body through food.

Since the 1780s, and still today, some of its compounds (i.e., bismuth subnitrate or bismuth subcarbonate, or bismuth subcitrate) are used for the treatment of gastric disorders, especially peptic ulcers.

Excess bismuth can cause liver damage.

Certain bismuth compounds are used as catalysts in manufacturing the acrylonitrile monomer, and also in the production of bismuthoxychloride, whenever a pearl effect is needed (i.e., in nail varnish and lipstick).

BISPHENOL-A ($C_{15}H_{16}O_2$)

Alternative Names/Abbreviations: 4,4'-(Propan-2-ylidene)diphenol, p,p'-isopropylidenebispheno 4,4'-dihydroxy-2,2-diphenylpropane, BPA

CAS Registry Number: 80-05-7

Details: A solid in flake, crystal or dust form. Used in production of epoxy-phenolic resins, monomer of polycarbonates (PC), an antioxidant for PVC, and as an inhibitor used during PVC polymerisation. PC are widely used in many consumer products, from sunglasses and CD to water and food containers and shatter-resistant baby bottles. Some polymers can also contain bisphenol A, and epoxy resins containing bisphenol A are common coatings used in food cans.

Toxicology: Bisphenol A can be released during thermal decomposition of epoxides above 250 °C, which can cause 'photosensitisation' of the skin. Bisphenol A is not genotoxic *in vivo*, however, weak oestrogenic effects are observed *in vitro*. Some endocrine disruptor effects in animals and human cancer cells can occur at low levels (2-5 ppb). It has been claimed that these effects lead to health problems, mainly in men, of a lowered sperm count and infertile sperm.

No TLV-TWA value is available yet for bisphenol A.

BORON (B)

CAS Registry Number: 7440-42-8

Boron is an essential element for plants, and humans take about 2 mg daily through vegetables and fruits. The total amount of boron in the body is about 18 mg. Five grams of boric acid can make a person ill, and 20 g or more can put life in danger.

Sodium borate (or borax) is well known as a skin ointment (along with lead, mercury and sulfur).

Boron compounds are used in making glass and detergents as well as used in agriculture. Boron is used in fibreglass and borosilicate fibre production, both used in fibre-reinforced composite systems. They are also used in fibreproofing polymeric fabrics.

Glass fibre used in reinforcing plastics and as insulation in buildings is a borosilicate.

Ceramic glazes for tiles and kitchen equipment are made of boric acid.

Boron compounds are also used as food preservatives, and boron compounds such as sodium octaborate are used to fireproof fabrics. Boric acid is also used as insecticide, in particular for ants and cockroaches in homes.

Compounds of borate are used as fertilisers.

Boron nitrade (microfine) powders are used in face powders to add a lustre and a silky feel to it.

BROMINE (Br$_2$) AND ITS COMPOUNDS

CAS Registry Number: 7726-95-6

Details and Toxicology: Bromine is a dark, reddish-brown, volatile liquid. It is used in the production of flame retardants for plastics. Bromine is a strong irritant to eyes, mucous membranes, lungs and skin. It only takes a concentration of 10 ppm to cause irritation in the upper airways, 40-60 ppm can easily cause pulmonary oedema, and 1000 ppm can be fatal. Exposure can be oral, by inhalation or by skin absorption.

TLV-TWA value for bromine is by ACGIH, 2003: 0.1 ppm (corresponding to 0.66 mg/m^3).

Bromodichloromethane (CAS Registry Number: 75-27-4) is a colourless liquid used as a solvent and as a flame retardant, and *polybrominated diphenylether* is used as flame retardant for furniture foams, plastic casings and some textiles, but since they can act as hormone mimics, these are seen as a threat. Bromodichloromethane is a central nervous system depressant, and can damage kidneys and liver. Exposure to bromodichloromethane is usually by inhalation or by skin absorption.

IARC classifies bromodichloromethane as a possible carcinogenic agent to humans.

1,3-BUTADIENE (C$_4$H$_6$, SMILES C=CC=C)

Alternative Names/Abbreviations: Butadiene, biethylene, divinyl, erythrene, vinylethylene

CAS Registry Number: 106-99-0

Details: 1,3-Butadiene is a colourless gas, it is the monomer of rubber (i.e., styrene-butadiene, polybutadiene rubber, as well as neoprene rubber). The word butadiene, most of the time, refers to 1,3-butadiene.

Toxicology: Exposure is usually by inhalation. 1,3-Butadiene is an irritant (for eyes and mucous membranes) and a probable A2 carcinogenic agent for humans (IARC). At acute high exposures, damage to the central nervous system can occur, and there are studies showing butadiene exposure also increases risks of cardiovascular diseases.

TLV-TWA value for 1,3-butadiene is by ACGIH, 2003: 2 ppm (corresponding to 4.4 mg/m^3).

n-BUTYL ACETATE ($C_6H_{12}O_2$), SMILES CCCCO(CO)C]

Alternative Names/Abbreviations: 1-Butyl acetate, 1-butyl ethanoate, butyl, butyl ethanoate, acetic acid butyl ester

CAS Registry Number: 123-86-4

Details: A colourless liquid, a solvent and a chemical used for lacquer and polymer production.

Toxicology: Exposure is usually by inhalation, it is an irritant (eyes and mucous membranes), a narcotic and a neurotoxic.

TLV-TWA value for *n*-butyl acetate is by ACGIH, 2003: 150 ppm (corresponding to 713 mg/m³), with a TLV-STEL value of 200 ppm (corresponding to 950 mg/m³).

(The data for *sec*-butyl acetate and *tert*-butyl acetate is similar to that given for *n*-butyl acetate).

n-BUTYL ACRYLATE ($C_7H_{12}O_2$)

Alternative Names/Abbreviations: Acrylic acid butyl ester, 2-propeonic acid butyl ester

CAS Registry Number: 141-32-2

Details: A colourless liquid used to produce acrylic polymers and fibres, used for paints.

Toxicology: Exposure is by inhalation or by skin contact. *n*-Butylacrylate is a strong irritant (to eyes and skin). There is no available data for its carcinogenic effect to humans (IARC), although it is a known carcinogen for animals.

TLV-TWA value for *n*-butylacrylate is by ACGIH, 2003: 10 ppm (corresponding to 52 mg/m³).

n-BUTYL ALCOHOL ($C_4H_{10}O$)

Alternative Names/Abbreviations: n-Butanol, butyric alcohol, propyl carbinol, butyl hydroxide, 1-butanol

CAS Registry Number: 71-36-3

Details: A colourless liquid used as a solvent, in the synthesis of polymers and rubber cements.

Toxicology: Exposure is by inhalation or by skin contact. It is an irritant (eyes and mucous membranes), a narcotic and a neurotoxic.

TLV-C value for *n*-butyl alcohol is by ACGIH, 2003: 50 ppm (corresponding to 152 mg/m^3), with a notation for skin absorption.

(The data for *sec*-butyl acetate and *tert*-butyl acetate is similar to that given for *n*-butyl acetate).

BUTYLAMINE (CH$_3$[CH$_2$]$_3$NH$_2$)

Alternative Names/Abbreviations: 1-Amino butane, n-butyl amine

CAS Registry Number: 109-73-9

Details: A colourless liquid, intermediate in the production of rubber.

Toxicology: Exposure is by inhalation or by skin contact, an irritant (eyes and mucous membranes and skin).

TLV-C value for butylamine is by ACGIH, 2003: 5 ppm (corresponding to 15 mg/m^3), with a notation for skin absorption.

BUTYLATED HYDROXYTOLUENE (C$_{15}$H$_{24}$O)

Alternative Names/Abbreviations: Butylated hydroxytoluene (BHT); 2,6-bis (1,1-dimethylethyl)-4-methylphenol, BHT

CAS Registry Numbers: 128-37-0

Details: A white, crystalline solid. BHT is used as an antioxidant in rubber and plastics.

Toxicology: Exposure is by ingestion. It has a low acute toxicity, and the IARC say that there is limited evidence for its carcinogenic effect on animals.

TLV-TWA value for BHT is by ACGIH, 2003: 2 mg/m^3.

Tert-BUTYL CHROMATE ($C_8H_{18}CrO_4$)

Alternative Names/Abbreviations: Bis (tert-butyl) chromate, chromic acid, di-tert-butyl ester

CAS Registry Number: 1189-85-1

Details: A colourless liquid, used as polymerisation catalysts for polyolefins and as a curing agent for urethanes.

Toxicology: Exposure is by inhalation or by skin contact, it is an irritant (eyes, nose and skin), and an inferred carcinogen.

TLV-C value for butyl chromate is by ACGIH, 2003: 0.1 ppm (as chromium trioxide) with a notation for skin absorption.

p-tert-BUTYL TOLUENE ($(CH_3)_3CC_6H_4CH_3$)

Alternative Names/Abbreviations p-Methyl-tert-butylbenzene, 1-methyl-4-tert butylbenzene, TBT

CAS Registry Number: 98-51-1

Details: A colourless liquid, used as a solvent for resins.

Toxicology: Exposure is by inhalation, it is an irritant (mucous membranes) and a central nervous system depressant. Can cause damage to lungs, brain, liver and kidneys in animals. Its characteristic odour is traceable at 5 ppm.

TLV-TWA value for *p-tert*-butyl toluene is: 1 ppm (corresponding to 6.1 mg/m^3).

CADMIUM (Cd) AND ITS COMPOUNDS

CAS Registry Number: 7440-43-9

Details: Metallic cadmium is a soft, silvery and ductile solid.

Cadmium oxide can be a colourless, amorphous powder or a red/brown crystal. The natural human intake of cadmium is between 10 to 1000 mg whereas WHO set the safe daily intake as 70 mg.

Toxicology: *Metallic cadmium* is an accumulative poison and it is in the UN Environmental Programmes list of top 10 hazardous pollutants. Cadmium mimics the element zinc which is essential to the body, and cadmium is absorbed by the body easily, where it prefers to accumulate in the kidneys, up to the point when they can no longer cope with it. When the level exceeds 200 ppm, it begins to prevent reabsorption of proteins, glucose and amino acids, damaging the filtering system and thus leading to kidney failure. Once in the human body, cadmium can remain for 30 years. Zinc is part of DNA polymerase, which is crucial to the production of sperm, its replacement by cadmium can be particularly damaging to the testicles. Excess exposure of cadmium can lead to weakening the bones and joints, making movement painful (the Itai/Itai disease in Japan). There are concerns that cadmium causes cancer in humans. Exposure to cadmium is by inhalation (most dangerous) or by ingestion.

Cadmium oxide fumes and *cadmium dust* are strong pulmonary irritants, the latter having a much stronger effect. Both are nephrotoxic and A2 suspected carcinogenic for humans (IARC) [14].

Cadmium sulfide is a common pigment (the bright cadmium yellow), and although it has been banned for some time, it was used for a long time in paints, artist's colours, plastics, rubbers, printing inks and vitreous enamels.

Cadmium red is a red pigment used for containers, toys, household items and it replaced pigments derived from toxic heavy metals such as lead and mercury, but cadmium is now considered environmentally undesirable and its pigments are being phased out, and replaced with cerium sulfide, which gives a rich red colour and is stable up to 350 °C.

TLV-TWA value for cadmium and its compounds is by ACGIH, 2003: 0.01 mg/m^3 for total particulates (or 0.002 mg/m^3 for the respirable fraction of dust).

CAMPHOR (C$_{10}$H$_{16}$O, SMILES CC1(C)C2(C)C(CC1CC2)=O)

Alternative Names/Abbreviations: 2-Bornanone, 2-keto 1,7,7, trimethylnorcamphane, bornan-2-one, Formosa camphor

CAS Registration Number: 76-22-2

Details: Crystals, with a special odour; a plasticiser specifically for cellulosics, lacquers and varnishes. Camphor, 464-49-3 ((1*R*)-Camphor), 464-48-2 ((1*S*)-Camphor)

Toxicology: Exposure is by inhalation or by skin contact. It is an irritant (eyes and nose), but not a suspected carcinogen. In large quantities, it is poisonous when ingested, and can cause seizures, confusion, and neuromuscular hyperactivity. In 1980, the FDA set the limit of a 11% maximum of cadmium in consumer products. Since alternative treatments exist, medicinal use of camphor is discouraged by the FDA, except for skin-related uses, such as medicated powders, which contain only small amounts of camphor.

TLV-TWA value for camphor is by ACGIH, 2003: 2 ppm (corresponding to 12 mg/m^3).

CAPROLACTAM (C₆H₁₁NO)

Alternative Names/Abbreviations: ε-Caprolactam, 2-oxohexamethylenimine, aminocaproic acid lactam

CAS Registry Number: 105-60-2

Details: A crystalline solid, the monomer for poly-caprolactam (Nylon 6).

Toxicology: Exposure is by inhalation. It is an irritant (eyes, mucous membranes, upper airways, and skin). The IARC concluded that there is no solid evidence for caprolactam's supposed carcinogenic effect on animals and humans. Caprolactam is mostly considered as an irritant and is toxic by ingestion, inhalation, or absorption through the skin.

TLV-TWA value for caprolactam (dust) is by ACGIH, 2003: 3 mg/m³.

CARBON BLACK

Alternative Names/Abbreviations: Activated carbon, acetylene carbon, decolourising carbon, actibon, ultra carbon, channel black, furnace black, thermal black, gas black, lamp black

CAS Registry Number: 1333-86-4

Details: Black crystal/amorphous powder, used in the processing of rubber and plastics as a reinforcing agent, as well as in paints. Carbon black is produced by the incomplete combustion of petroleum products. It has an extremely high surface to volume ratio, and it is one of the first nanomaterials to be commonly used.

Toxicology: Exposure is by inhalation. The IARC determined that carbon black extract is a carcinogen for animals, but there is no health hazard documented for humans [15]. However, carbon black is a suspected carcinogen and is specifically harmful to the upper respiratory tract if inhaled, because of its large particulate content.

TLV-TWA value for carbon black is by ACGIH, 2003: 3.5 mg/m³.

CATECHOL (C₆H₄(OH)₂)

Alternative Names/Abbreviations: 1,2-dihydroxybenzene, pyrocatechol, 1,2-benzenediol, pyrocatechin

CAS Registry Number: 120-80-9

Details: Crystalline, used in the production of rubber, antioxidants and inhibitors, but mainly as a film developing chemical.

Toxicology: Exposure is by inhalation or by skin contact. Catechol is an irritant (eyes, mucous membranes, upper airways, and skin). The IARC concluded that catechol is a carcinogen for animals, and a possible carcinogen for humans [16].

TLV-TWA value for catechol is by ACGIH, 2003: 5 ppm (corresponding to 20 mg/m^3), with a notation for skin absorption.

CHLORINATED DIBENZO-p-DIOXINS (C$_{12}$H$_4$Cl$_4$O$_2$)

Alternative Names/Abbreviations: a general name given to a group of 75 different polychlorinated dibenzo-p-dioxins (PCDD), dichlorinated dioxins (DCDD), trichlorinated dioxins (TrCDD), tetrachlorinated dioxins (TCDD), etc

CAS Registry Number: 136677-09-3

Details: They can be solids or liquids. They are produced during the production, and from incomplete combustion, of certain chlorinated organic compounds, and also produced naturally during certain human activities, as well as by forest fires and volcanic eruptions. One member of the group is the herbicide, 'Agent Orange' (used as a chemical weapon during the Vietnam War), and also used as a wood preservative.

The Seveso disaster in Italy in 1976 was mainly as a result of dioxins and affected a number of people in around the town Seveso.

Toxicology: Dioxins are a group of persistent organic pollutants. Being lipophilic, they bioaccumulate easily in fatty tissues. Some of them have been proven to be extremely toxic to animals, but scientific evidence of harmfulness to humans is still disputed. A subgroup of PCDD are amongst the most toxic. The number and position of the chlorine atoms determine the extent of toxicity of the PCDD, and the most toxic among them is 2,3,7,8-tetrachlorodibenzo-*p*-dioxin (commonly referred to as 'dioxin'). Exposure is by inhalation or by skin contact. They are highly toxic (to the immuno-reproductive-developmental and nervous systems) and are a suspected human carcinogen, the most severely effective one is TCDD. According to the IARC, 2,3,7,8-CDD is a Group 1 human carcinogen [17].

The Toxicity Equivalent Factor (TEF Method), developed and validated in animals, is expressed as 1.0 for TCDD and for all other dioxin-like PCB it is less than 1.0.

β-CHLOROPRENE (C$_4$H$_5$Cl) SMILES C=C(Cl)C=C

Alternative Names/Abbreviations: Chlorobutadiene, 2-chloro-1,3-butadiene, chloroprene

CAS Registry Number: 126-99-8

Details: A liquid, used in the production of the rubber polychloroprene (known as Neoprene).

Toxicology: Exposure is by inhalation or by skin contact. A strong irritant (eyes and skin), a toxin (reproductive, mutagenic), as well as a proved carcinogen to animals and a possible carcinogen to humans [18].

TLV-TWA value for β-chloroprene is by ACGIH, 2003: 10 ppm (corresponding to 36 mg/m^3), with a notation for skin absorption.

α-CHLOROSTYRENE (C$_8$H$_7$Cl)

Alternative Names/Abbreviations: 2-Chlorostyrene

CAS Registry Number: 2039-87-4

Details: A liquid monomer used for the production of some specialty polymers.

Toxicology: Exposure is by inhalation. It is an irritant to eyes, nose and to mucous membranes, and is toxic to the central nervous system. No adverse effects are known for humans, but it causes the kidneys and liver to increase in size.

TLV-TWA value for α-chlorostyrene is by ACGIH, 2003: 50 ppm (corresponding to 283 mg/m^3).

CHROMIUM (Cr) AND ITS COMPOUNDS

CAS Registry Number: 18540-29-9 (Cr Compounds)

Details: A solid metal, which can have valences of 2, 3 or 6 in its compounds. They are used as catalysts.

Toxicology: *Metallic chromium and its 2- and 3-valent compounds*: exposure is by inhalation. According to the IARC, there is insufficient evidence for the carcinogenicity of metallic chromium, as well as its 2- and 3- valent compounds.

Hexavalent chromium compounds: exposure is by inhalation. Water soluble hexavalent chromium compounds (such as chromic acid mists and chromate dusts) are severe irritants (lungs and skin), whereas water insoluble ones are confirmed A1 carcinogenic agents to humans [19].

TLV-TWA value for chromium (VI) compounds are by ACGIH, 2003: 0.01 mg/m^3.

COBALT (Co) AND ITS COMPOUNDS

CAS Registry Number: 7440-48-4

Details: Cobalt is a grey solid material. It is used mainly in paints.

Toxicology: Exposure is by inhalation. Cobalt causes skin allergies and irritation, and its dust causes occupational asthma. The IARC decided that cobalt metal powder and cobaltous oxide are A3 confirmed carcinogenics in animals, and they are possible human carcinogens [20].

TLV-TWA value for cobalt and its compounds are by ACGIH, 2003: 0.02 mg/m^3.

CRESOL (ALL ISOMERS) (C_7H_8O)

Alternative Names/Abbreviations: Ortho/meta/para-cresols, cresylic acid, tricresol, methyl-phenol, hydroxytoluene

CAS Registry Numbers: 95-48-7 for *ortho* 108-39-4 for *meta* 106-44-5 for *para* 1319-77-3 for their mixtures

Details: *Meta*-cresol and its isomer mixtures are liquids, all others are solids. They are used as flame retardant plasticisers and wood preservatives.

Toxicology: Exposures at very low levels are believed to be not harmful. When breathed, ingested, or applied to the skin at very high levels, they can be very harmful, resulting in irritation of the nose and throat, burning of skin, eyes, mouth, and throat, abdominal pain and vomiting, heart damage, anaemia, liver and kidney damage, disturbance of the central nervous system, and acting as a weak genotoxic agent, causing facial paralysis and finally coma and death. However, little is known about the effects of breathing cresols at much lower levels over longer periods of time [21]. Its odour is distinct at low concentrations (5 ppm). In general, *ortho* and *para* isomers are the most toxic with equal toxicities.

TLV-TWA value for cresols is by ACGIH, 2003: 5 ppm (corresponding to 22 mg/m^3), with a notation for skin absorption.

CYANAMIDE (CH_2N_2)

Alternative Names/Abbreviations: Carbodiimide, cyano-amine, hydrogen cyanamide, cyanogen nitride

CAS Registry Number: 420-04-2

Details: Cyanamide is a crystalline solid, used for the production of synthetic rubber.

Toxicology: Exposure is by inhalation and ingestion. Cyanamide is an irritant (eyes, mucous membranes, upper airways and skin), and has strong toxic effects.

TLV-TWA value for cyanamide is by ACGIH, 2003: 2 mg/m^3.

CYCLOHEXANE (C_6H_{12})

Alternative Names/Abbreviations: Hexahydrobenzene, benzene hexahydride, hexamethylene

CAS Registry Number: 110-82-7

Details: A colourless liquid, which is used to produce intermediates of polyamides, and is a solvent for resins and rubber.

Toxicology: Exposure is by inhalation. Cyclohexane is an irritant (eyes and mucous membranes), and a narcotic at high concentrations.

TLV-TWA value for cyclohexane is by ACGIH, 2003: 300 ppm (corresponding to 1030 mg/m^3).

CYCLOHEXANOL ($C_6H_{11}OH$)

Alternative Names/Abbreviations: Hexahydrophenol, cyclohexyl alcohol

CAS Registry Number: 108-93-0

Details: A colourless liquid, and a solvent for most of the resins used in the production of polymers

Toxicology: Exposure is by inhalation and by skin absorption. Cyclohexanol is an irritant (eyes and mucous membranes), and a narcotic at high concentrations.

TLV-TWA value for cyclohexanol is by ACGIH, 2003: 50 ppm (corresponding to 206 mg/m^3), with a notation for skin absorption.

CYCLOHEXANONE ($C_6H_{10}O$)

Alternative Names/Abbreviations: Pimelic ketone, hexanon, sextone

CAS Registry Number: 108-94-1

Details: A colourless liquid with a peppermint odour, and a solvent for most of the resins and rubber, it is used in the production of polymers.

Toxicology: Exposure is by inhalation and by skin absorption. Cyclohexanone is an irritant (eyes and mucous membranes), and a narcotic at high concentrations.

TLV-TWA value for cyclohexanone is by ACGIH, 2003: 25 ppm (corresponding to 100 mg/m^3), with a notation for skin absorption.

DIACETONE ALCOHOL ($C_6H_{12}O_2$ or $(CH_3)_2C(OH)CH_2COCH_3$)

Alternative Names/Abbreviations: 4-Hydroxy-4-methyl-2-pentanone, diacetone, dimethyl acetonyl carbinol

CAS Registry Number: 123-42-2

Details: A liquid, and a solvent for most of the resins and pigments, it is used in the production of polymers.

Toxicology: Exposure is by inhalation and by skin absorption (secondary). Diacetone alcohol is an irritant (eyes and mucous membranes), and a narcotic at high concentrations. Its odour threshold is rather low: 0.3 ppm.

TLV-TWA value for diacetone alcohol is by ACGIH, 2003: 50 ppm (corresponding to 238 mg/m^3).

2,4-DIAMINOTOLUENE ($C_7H_{10}N_2$)

Alternative Names/Abbreviations: Toluene, 2,4-diamine, 3-amino-p-toluidine, 1,3-diamino-4-methyl benzene, TDA, 2,4-TDA

CAS Registry Number: 95-80-7

Details: A crystalline solid, and an intermediate for toluene diisocyanate used for production of polyurethanes.

Toxicology: Exposure is by inhalation and by skin absorption. Diaminotoluene is an irritant (eyes and skin), and a proved carcinogen and a reproductive toxin for animals.

There is no available TLV-value for diaminotoluene.

DIBORANE (B_2H_6)

Alternative Names/Abbreviations: Boroethane, boron hydride

CAS Registry Number: 19287-45-7

Details: A gas, an initiator for polymerisations (ethylene, vinyl and styrene).

Toxicology: Exposure is by inhalation. Diborane is an irritant (pulmonary system), and the threshold odour detection rotten eggs is 3.3 ppm.

TLV-TWA value for diborane is by ACGIH, 2003: 0.1 ppm (corresponding to 0.11 mg/m^3).

DIBUTYL PHTHALATE (C$_{16}$H$_{22}$O$_4$)

Alternative Names/Abbreviations: Butyl phthalate, 1,2-benzenedicarboxylic acid, dibutylester, phthalic acid dibutyl ester, DBP

CAS Registry Number: 84-74-2

Details: It is a liquid, one of the common plasticisers for plastics (PVC).

Toxicology: Exposure is by inhalation and by ingestion, with low level acute toxicity [22]. DBP was added to the California Proposition 65 (1986) List of Suspected Teratogens in November 2006.

TLV-TWA value for DBP is by ACGIH, 2003: 5 mg/m^3.

DICHLOROACETYLENE (Cl$_2$C$_2$)

Alternative Names/Abbreviations: Dichloroethyne

CAS Registry Number: 7572-29-4

Details: A liquid, a by-product in the synthesis of the monomer vinylidene chloride.

Toxicology: Exposure is by inhalation. Dichloroacetylene is a neurotoxin and an A3 carcinogen for animals. The IARC does not list it as a carcinogen for humans.

TLV-C value for dichloroacetylene is by ACGIH, 2003: 0.1 ppm (corresponding to 0.39 mg/m^3).

DICHLORODIFLUOROMETHANE (CCl$_2$F$_2$), SMILES ClC(F)(Cl)F

Alternative Names/Abbreviations: Freon/Refrigerant 12, isotron, CFC-12

CAS Registry Number: 75-71-8

Details: A gas, used in the recent past as a blowing agent for cellular polymer production.

Toxicology: Exposure is by inhalation. Dichlorodifluoromethane is a neurotoxin.

TLV-TWA value for dichlorodifluoromethane is by ACGIH, 2003: 1000 ppm (corresponding to 4950 mg/m^3).

1,3-DICHLORO-5,5-DIMETHYLHYDANTOIN ($C_5H_6Cl_2O_2N_2$)

Alternative Names/Abbreviations: Dactin, Halane, DCDMH

CAS Registry Number: 118-52-5

Details: A white, powdery solid, used as a catalyst for polymerisation reactions.

Toxicology: Exposure is by inhalation. DCDMH is an irritant (eyes and mucous membranes) and a neurotoxin.

TLV-TWA value for DCDMH is by ACGIH, 2003: 0.2 mg/m^3.

DIEPOXYBUTANE ($C_4H_6O_2$)

Alternative Names/Abbreviations: 2,2´-Bioxirane, 1,1´-bi(ethylene oxide), butadiene diepoxide, butadiene dioxide, 2,4-diepoxy butane, dioxybutadiene

CAS Registry Number: 1464-53-5

Details: A liquid, used as a curing agent for polymers, and as a biocide.

Toxicology: Exposure is by inhalation, and skin absorption. Diepoxybutane is an irritant (mucous membranes), and the IARC. It is considered a carcinogenic agent for animals.

No TLV value is available.

DIETHYLAMINE [(C_2H_5)$_2$NH, SMILES CCNCC]

Alternative Names/Abbreviations: Diethylamine, N-ethylethanamine

CAS Registry Number: 109-89-7

Details: A liquid; used in the production of polymers. It has a specific ammonia odour which is detectable at 0.13 ppm.

Toxicology: Exposure is by inhalation. Diethylamine is an irritant (eyes, skin and mucous membranes), with an IARC classification of A4 not classifiable as a human carcinogen.

TLV-TWA value for diethylamine is by ACGIH, 2003: 5 ppm (corresponding to 15 mg/m^3).

DIETHYLENETRIAMINE [C$_4$H$_{13}$N$_3$, SMILES NCCNCCN]

Alternative Names/Abbreviations: 2,2-Diaminodiethylamine, DETA

CAS Registry Number: 111-40-0

Details: A liquid with an ammonia-like odour, used as a hardener and stabiliser for polymers (specifically for epoxy resins).

Toxicology: Exposure is by inhalation and skin contact. DETA is an irritant (skin, eyes, upper airways).

TLV-TWA value for DETA is by ACGIH, 2003: 1 ppm (corresponding to 4.2 mg/m^3), with a notation for skin absorption [23].

DIMETHYLHEXYL ADIPATE (C$_{22}$H$_{42}$O$_4$)

Alternative Names/Abbreviations: Bis-[2-ethylhexyl] adipate, octyl adipate, DMHA

CAS Registration Numbers: 103-23-1

Details: A liquid, used as a plasticiser and plastisol in PVC.

Toxicology: Exposure is by inhalation and skin contact. DMHA has a low acute toxicity, and is a carcinogen for animals. The IARC does not classify DMHA as a human carcinogen.

No established TLV value is available for DMHA.

DI[2-ETHYLHEXYL] PHTHALATE (C$_{24}$H$_{38}$O$_4$)

Alternative Names/Abbreviations: Bis(2-ethylhexyl) phthalate, diethylhexyl phthalate, di-sec-octyl phthalate, DEHP

CAS Registration Number: 117-81-7

Details: An oily liquid, used commonly as a plasticiser and plastisol in PVC (specifically for the preparation of PVC plastic films for biomedical and food packaging applications).

Toxicology: Exposure is mostly by inhalation. DEHP has very low acute (lethal) toxicity.

DEHP can affect certain organs (mostly reproductive organs, lungs, kidney, and liver. Even at low concentrations, it can be embryotoxic, and can also be found in breast milk. The IARC determined that there is sufficient evidence for the carcinogenicity of DEHP in certain animals, and inadequate evidence in humans, hence, it is classified as A3 - confirmed animal carcinogen with unknown relevance to humans [24, 25]. Oral DEHP is hydrolysed to mono-ethylhexyl phthalate, although its metabolism may be different for different conditions.

TLV-TWA value for DEHP is by ACGIH, 2003 and OSHA): 5 mg/m^3 of air. The EPA limit for DEHP in drinking water is 6 ppb.

DIGLYCIDYL ETHER ($C_6H_{10}O_3$)

Alternative Names/Abbreviations: Bis (2,3-epoxy propyl) ether, di (2,3-epoxypropyl) ether, DGE

CAS Registry Number: 2238-07-5

Details: A liquid, used as a diluent for epoxy resins and as a stabiliser.

Toxicology: Exposure is by inhalation and skin contact. DGE is a strong irritant (eyes, upper airways and skin), and a strong toxic and mutagenic agent [26].

TLV-TWA value for DGE is by ACGIH, 2003: 0.1 ppm (corresponding to 0.53 mg/m^3).

DIISOBUTYLPHTHALATE ($C_{16}H_{22}O_4$)

Alternative Names/Abbreviations: Phthalic acid diisobutyl ester, DIBP

CAS Registry Number: 84-69-5

Details: DIBP is one of the rare odourless plasticisers with excellent heat/light stabilities, and it is inexpensive. DIBP is a commonly used as a plasticiser for cellulose nitrate production. DIBP has not yet undergone an EU classification and labelling assessment and it is very unlikely that it has any toxicological effects. Since it is not known for sure, its toxicology information is not given here (http://www.dibp-facts.com/index.asp?page=3).

DIMETHOXYETHYL PHTHALATE $C_6H_4(COOCH_2\ CH_2OCH_3)_2$

Alternative Numbers/Abbreviations: 1,2-Benzenedicarboxylic acid, bis(2-methoxy-ethyl) phthalate, bis(methoxyethyl) phthalate, dimethyl cellosolve phthalate, DMEP

CAS Registry Number: 117-82-8

Details: A liquid, used as a solvent and a plasticiser for plastics.

Toxicology: Exposure is by inhalation. DMEP has slight to moderate toxicity, with reproductive effects, in animals.

No established TLV value is available for DMEP.

N,N-*DIMETHYLANILINE* C$_6$H$_5$N(CH$_3$)$_2$

Alternative Names/Abbreviations: Dimethylphenylamine, aminodimethylbenzene, dimethylaniline

CAS Registry Number: 121-69-7

Details: A liquid, used as a solvent for plastics, and as a vulcanising agent and stabiliser for rubbers.

Toxicology: Exposure is by inhalation and skin contact. The IARC concluded that there is limited evidence in animals and inadequate evidence in humans as regards the carcinogenicity of dimethylaniline, hence it is not classified as a carcinogen for humans [27].

TLV-TWA value for N,N-dimethylaniline is by ACGIH, 2003: 5 ppm (corresponding to 25 mg/m^3).

DIMETHYL HYDROGEN PHOSPHITE (CH$_3$O)$_2$POH

Alternative Names/Abbreviations: Dimethoxyphosphine oxide, dimethyl phosphite, methyl phosphonate, DMHP

CAS Registration Number: 868-85-9

Details: A liquid, used as a flame retardant (specifically for Nylon 6 fibres)

Toxicology: Exposure is by inhalation. DMHP is an irritant (eyes, mucous membranes and skin), and a neurological toxin. The IARC concluded that there is limited evidence for the carcinogenicity of DMHP in animals and that it is not classifiable as carcinogenicity to humans [28].

No established TLV value is available for DMHP.

DIMETHYL METHYLPHOSPHONATE (C$_3$H$_9$O$_3$P)

Alternative Names/Abbreviations: Phosphonic acid methyl-dimethyl ester, DMMP

CAS Registration Number: 756-79-6

Details: A solid, used as a flame retardant, antistatic agent, antifoam agent, and as a plasticiser and stabiliser.

Toxicology: Exposure is by inhalation. DMMP is a reproductive toxicant and a carcinogen for animals. Its effects on humans are not known.

No established TLV value is available for DMMP.

DIMETHYL PHTHALATE (C$_{10}$H$_{10}$O$_4$)

Alternative Name/Abbreviation: Methyl phthalate, phthalic acid dimethyl ester, 1,2-benzenedicarboxylic acid dimethyl ester, DMPh

CAS Registration Number: 131-11-3

Details: A liquid, used as a plasticiser

Toxicology: Exposure is by inhalation and by skin. DMPh is a low-order acute toxic chemical.

TLV-TWA value for DMPh is by ACGIH, 2003: 5 mg/m^3 [23, 29, 30].

DINITRO-o-CRESOL CH$_3$C$_6$H$_2$OH(NO$_2$)$_2$

Alternative Name/Abbreviation: Dinitrol, 2-methyl-4,6-dinitrophenol, DNOC

CAS Registration Number: 534-52-1

Details: A yellow coloured crystalline sold, used as the polymerisation inhibitor for vinyl aromatic compounds.

Toxicology: Exposure is by inhalation, ingestion and by skin.

TLV-TWA value for DNOC is by ACGIH, 2003: 0.2 mg/m^3 with a notation for skin absorption [23, 29, 30].

DIOXANE ($C_4H_8O_2$)

Alternative Name/Abbreviation: 1,4-Dioxane, dioxyethylene ether, diethylene ether, 1,4-diethylene dioxide, 1,4-dioxycyclohexane, 1,4-dioxane, p-dioxane

CAS Registration Number: 123-91-1

Details: A liquid, used as a solvent in the manufacture of polymers, fibres and adhesives; and a stabiliser when used with chlorinated solvents.

Toxicology: Exposure is by inhalation and by skin. Dioxane is an irritant (eyes, and skin) and a carcinogen to animals. It has a low odour threshold (3-6 ppm). The IARC have classified dioxane as a carcinogen for animals and as group 2B possibly carcinogenic for humans [31].

TLV-TWA value for dioxane is by ACGIH, 2003: 25 ppm (corresponding to 90 mg/m^3), with a notation for skin absorption.

DIPHENYLAMINE ($(C_6H_5)_2NH$)

Alternative Name/Abbreviations: N-Phenylbenzeneamine, N-phenylaniline, N-diphenylaniline, N,N-diphenylamine, DPA

CAS Registry Number: 122-39-4

Details: A solid, used as an antioxidant and accelerator (rubber).

Toxicology: DPA is toxic. Exposure is by inhalation and by ingestion. DPA is known to damage kidneys and liver in animals [23, 29, 30].

TLV-TWA value for DPA is by ACGIH, 2003: 10 mg/m^3).

DIPROPYLENE GLYCOL METHYL ETHER ($CH_3OC_3H_6OC_3H_6OH$)

Alternative Names/Abbreviations: Dipropylene glycol monomethyl ether, DPGME

CAS Registration Number: 34590-94-8

Details: A liquid, a solvent for synthetics, a narcotic, and a neurotoxic.

Toxicology: Exposure is by inhalation. DPGME is an irritant (eyes).

TLV-TWA value for DPGME is by ACGIH, 2003: 100 ppm (corresponding to 606 mg/m^3), with a notation for skin absorption [23, 29, 30].

DIPROPYL KETONE (CH$_3$CH$_2$CH$_2$)$_2$CO

Alternative Names/Abbreviations: Propyl ketone, butryone, 4-heptanone, heptan-4-one, DPK

CAS Registration Number: 123-19-3

Details: A liquid, and a solvent for polymers. DPK is a narcotic, and a neurotoxic.

Toxicology: Exposure is by inhalation. Although no adverse effects for humans are reported, it is known as an irritant of skin for animals [23, 29, 30].

TLV-TWA value for DPK is by ACGIH, 2003: 50 ppm (corresponding to 233 mg/m^3).

DISULFIRAM (C$_{10}$H$_{20}$N$_2$S$_4$)

Alternative Names/Abbreviations: Thiuram E, tetraethyl thiuram disulfide, antabuse, bis (diethylthiocarbamoyl disulfide), TETD

CAS Registration Number: 97-77-8

Details: A solid, an accelerator and vulcanising agent for rubber and a plasticiser (for Neoprene).

Toxicology: Exposure is by inhalation. TETD is a neurotoxic for the central nervous system and causes 'antabuse-alcohol' syndrome, if associated with alcohol.

TLV-TWA value for TETD is by ACGIH, 2003: 2 mg/m^3 [23, 29, 30].

DIVINYL BENZENE (C$_{10}$H$_{10}$)

Alternative Names/Abbreviations: Vinyl styrene, diethyl benzene, 1,4-divinyl benzene, DVB

CAS Registration Number: 1321-74-0

Details: A liquid, a crosslinking agent and a monomer for synthetic rubber production.

Toxicology: Exposure is by inhalation. DVB is a strong irritant (eyes, nose, mucous membranes and skin), and slightly genotoxic.

TLV-TWA value for DVB is by ACGIH, 2003: 10 ppm (corresponding to 53 mg/m^3) [24-26].

EPICHLOROHYDRIN (C_3H_5OCl), SMILES C1C2(O1)C2Cl)

Alternative Names/Abbreviations: Chloromethyloxirane, 3-chloro-1,2 propylene oxide

CAS Registration Number: 106-89-8

Details: A liquid, used in the production of epoxy and phenoxy resins.

Toxicology: Exposure is by inhalation and skin absorption. Epichlorohydrin is a poison and a very strong irritant (eye, skin and upper airways), and a mutagen causing genetic damage in animals. It is harmful if inhaled and it may cause delayed lung damage. May be fatal if swallowed. Causes burns on skin. The IARC concluded that, epichlorohydrin is a proven carcinogen for animals and it is probably an A2 suspected carcinogen for humans [32].

TLV-TWA value for epichlorohydrin is by ACGIH, 2003: 0.1 ppm (corresponding to 0.38 mg/m³).

ETHYLACRYLATE ($C_5H_8O_2$)

Alternative Names/Abbreviations: Acrylic acid ethyl ester, ethyl 2-propenoate

CAS Registry Number: 140-88-5

Details: A liquid, and a monomer.

Toxicology: Exposure is by inhalation. Ethyl acrylate is an irritant (eyes, skin and upper air ways). The IARC concluded that there is enough evidence to classify it as carcinogenic for animals, and as a possible carcinogen for humans. Ethyl acrylate is detectable at rather low concentrations (1 ppm).

TLV-TWA value for ethyl acrylate is by ACGIH, 2003: 5 ppm (corresponding to 21 mg/m³).

ETHYLAMINE ($C_2H_5NH_2$, SMILES CCN)

Alternative Names/Abbreviations: Monoethylamine, aminoethane, ethanamine

CAS Registry Numbers: 75-04-7

Details: A liquid with an ammonia like characteristic smell, and a stabiliser for rubber latex.

Toxicology: Exposure is by inhalation. Ethylamine is an irritant (eyes, mucous membranes and skin).

TLV-TWA value for ethyl emine is by ACGIH, 2003: 5 ppm (corresponding to 27.6 mg/m³).

ETHYLENE THIOUREA (C₃H₆N₂S)

Alternative Names/Abbreviations: Imidazoldinethione, 2-mercaptoimidazoline, 2-imidazole-2-thiol, ETU

CAS Registration Number: 96-45-7

Details: A solid, which is an accelerator and vulcanising agent for chloroprene rubber (Neoprene) and polyacrylate rubber.

Toxicology: Exposure is by inhalation. ETU is a proven animal carcinogen, and the IARC have determined that there is not enough evidence to judge its carcinogenicity to humans [33].

No established TLV value is available for ETU.

FORMALDEHYDE (HCHO, SMILES C=O)

Alternative Names/Abbreviations: Methanol, formic aldehyde, formalin, formol, oxomethane, oxymethylene, methylene oxide, methyl aldehyde

CAS Registry Number: 50-00-0

Details: A gaseous substance, with a pungent smell, formaldehyde is used in the manufacture of formaldehyde resins.

Toxicology: Exposure is by inhalation. Formaldehyde resins are used in many construction materials, including plywood, carpet, and spray-on insulating foams, and all these can slowly give off formaldehyde over time, hence formaldehyde is one of the more common indoor air pollutants. It is an irritant (eyes and upper air ways, as well as to skin, and mucous membranes, resulting in watery eyes). Formaldehyde is detectable at rather low concentrations (1 ppm). Above concentrations of 0.1 mg/kg in air, it begins to irritate the eyes and mucous membranes, resulting in headache and difficulty in breathing. Formaldehyde is easily soluble in water, and its high exposures can be potentially lethal. It is converted to formic acid in the body, which leads to a rise in blood pressure, hypothermia, followed by coma or death.

Formaldehyde is a genotoxic chemical. It is a carcinogen, however, there is an on-going complication at this point as US-EPA accepts it as an A2 suspected carcinogen, whereas IARC considers it as a known human carcinogen [34].

TLV-TWA value for formaldehyde is by ACGIH, 2003: 0.3 ppm (corresponding to 0.37 mg/m^3).

HEXAMETHYLENE DIISOCYANATE (C$_8$H$_{12}$N$_2$O$_2$, SMILES O=C=NCCCCCCN=C=O)

Alternative Names/Abbreviations: 1,6-Diisocyanato-hexane, 1,6-hexamethylene diisocyanate, HDI

CAS Registry Number: 822-06-0

Details: A liquid, and a crosslinking agent and hardener during the production of polyurethane.

Toxicology: Exposure is by inhalation. HDI is an irritant (eyes, mucous membranes and skin). HDI can lead to denaturation of proteins in the body and to the loss of enzyme function [35].

TLV-TWA value for HDI is very low by ACGIH, 2003: 0.005 ppm (corresponding to 0.034 mg/m^3).

HEXAMETHYL PHOSPHORAMIDE (C$_{16}$H$_{18}$N$_3$OP)

Alternative Names/Abbreviations: Hexamethyl phosphoric triamide, hexametapol, HMPA

CAS Registry Number: 680-31-9

Details: A liquid, and a solvent for polymers. HMPA is a polymerisation catalyst, a stabiliser (in polystyrene against thermal degradation), and a UV stabiliser (for polyvinyl and polyolefinic resins).

Toxicology: Exposure is by inhalation and by skin exposure. HMPA causes kidney damage, testicular atrophy, and is a proved A3 carcinogen for animals. According to IARC, HMPA is possibly carcinogenic to humans as well [36]. However, HMPA was recently anticipated to be classified as a human carcinogen, according to the eleventh edition of the report on carcinogens [37].

No established TLV value is available for HMPA.

2-HYDROXYPROPYL ACRYLATE ($C_6H_{10}O_3$)

Alternative Names/Abbreviation: Propylene glycol monoacrylate, 1,2-propanediol-1-acrylate, HPA

CAS Registry Number: 25584-83-2

Details: A liquid, and a monomer for production of thermoset systems.

Toxicology: Exposure is by inhalation and by skin absorption. HPA is an irritant (eyes, upper airways and skin).

TLV-TWA value for HPA is low by ACGIH, 2003: 0.5 ppm (corresponding to 2.8 mg/m³) [23, 29, 30].

ISOBUTANE [C_4H_{10}, SMILES C(C)CC]

Alternative Names/Abbreviations: Trimethyl methane, 2-methyl propane

CAS Registry Number: 75-28-5

Details: A gas, used during the production of polyurethane foams and resins.

Toxicology: Exposure is by inhalation. Isobutane is of low toxicity, except at very high concentrations when it can produce narcosis and cardiac failures. It does not irritate eyes or skin at normal concentrations.

TLV-TWA value for isobutane is by ACGIH, 2003: 800 ppm (corresponding to 1900 mg/m³) [23, 29, 30].

ISOCYANATES

Some Details: A family of highly reactive, low molecular weight chemicals, containing characteristic functional group of atoms –N=C=O, which are widely used in the manufacture of flexible/rigid PU foams, as well as manufacturing of fibres, coatings and elastomers. Spray-on PU products are used to protect wood, fibreglass, concrete and metals.

The most widely used isocyanate compounds are the **diisocyanates**, which have two isocyanate groups [i.e., methylene diisocyanate (MDI, CAS Registry Number: 822-06-0), toluene diisocyanate (TDI, CAS Registry Number: 584-84-9), hexamethylene diisocyanate (HDI, CAS Registry Number: 822-06-0), naphthalene diisocyanate (NDI, CAS Registry Number: 25554-28-4), methylene bis-cyclohexylisocyanate (CAS Registry Number: 5124-30-1), or hydrogenated MDI (HDI), and isophorone diisocyanate (IPDI, CAS Registry Number: 4098-71-9), as given next for additional information].

Polyisocyanates, have several isocyanate groups (i.e., HDI biuret and HDI isocyanurate) [23, 29, 30].

Toxicology: Isocyanates are very reactive which makes them harmful to living tissues. They are strong irritants to the mucous membranes and gastrointestinal and respiratory tracts, leading to asthma. Direct skin contact can cause severe inflammation.

ISOPHORONE DIISOCYANATE ($C_{12}H_{18}N_2O_2$)

Alternative Names/Abbreviations: 3-Isocyanatomethyl, 3,5,5-trimethyl cyclohexylisocyanate, IPDI

CAS Registry Number: 4098-71-9

Details: A liquid, used in polyurethane paints, casting compounds and coatings.

Toxicology: Exposure is by inhalation and by skin absorption. IPDI is an irritant (eyes, upper airways and skin). IPDI can cause bronchitis, asthma and allergic dermatitis.

TLV-TWA value for IPDI is low by ACGIH, 2003: 0.005 ppm (corresponding to 0.045 mg/m^3) [23, 29, 30].

N-ISOPROPYLANILINE ($C_9H_{13}N$)

Alternative Names/Abbreviations: Benzeneamine, N-phenylisopropylamine, NIPA

CAS Registry Number: 768-52-5

Details: A liquid, used in dyeing acrylics.

Toxicology: Exposure is by inhalation and by skin absorption.

TLV-TWA value for NIPA is by ACGIH, 2003: 5 ppm (corresponding to 12 mg/m^3 with the STEL value of 10 ppm [23, 29, 30], and NIOSH recommended exposure limit (REL): 2 ppm TWA (skin) [30].

ISOPROPYL GLYCIDYL ETHER ($C_6H_{12}O_2$)

Alternative Names/Abbreviations: 1,2-Epoxy-3-isopropoxypropane, IGE

Details: A liquid, used in production of epoxies, and a stabiliser.

Toxicology: Exposure is mainly by inhalation. IGE is an irritant (eyes, skin and mucous membranes), however, no systematic effect of IGE is observed for humans.

TLV-TWA value for IGE is by ACGIH, 2003: 50 ppm (corresponding to 238 mg/m³), with the STEL value of 75 ppm [23, 29, 30].

LEAD (Pb) AND COMPOUNDS (ORGANIC AND INORGANIC)
(Lead and Inorganic Lead compounds)

CAS Registry Number: 7439-92-1

Details: They are either water insoluble or soluble solids, the latter being more poisonous to humans.

Toxicology: Exposure is mainly by inhalation and ingestion.

Lead has no biological role and it acts as a cumulative poison. The most serious adverse effects: mental retardation and learning problems, occur in young children subjected to chronic exposure, most often through ingestion of paints. All forms of lead are extremely toxic to humans. Children with iron and calcium deficiencies absorb more lead and hence there is greater adverse effect. The main effect on adults is neurological. The initial symptoms of mild lead poising are headaches, nausea, stomach pains, vomiting, joint pain and conspitation. At higher exposure levels, there is toxic psychosis. It can cause hypertension, severe gastrointestinal problems and anaemia to neuromuscular dysfunctions, neurophysiological changes and kidney diseases, and neurological effects (especially on children), digestive problems, sterility, miscarriages, and possibly cancer. A single dose is unlikely to kill, but its absorption over a period of time is fatal. It is stored away in the bones as lead phosphate.

Lead compounds are preferentially concentrated in a cumulative way, in soft tissues after absorption, mainly in the liver and kidneys. Certain soluble inorganic lead compounds (i.e., lead acetate and lead phosphate) are A3 confirmed carcinogens to animals, and unknown effects on humans [38].

Lead chromate, a water-insoluble compound also called Chrome Yellow and used as a pigment, is an A2 suspected human lung carcinogen and can cause chronic lead poisoning (IARC [19]).

TLV-TWA value for lead and its inorganic compounds, as Pb is by ACGIH, 2003: 0.05 mg/m³).

Organolead compounds contain a direct chemical bond between carbon and lead. Their use is rather limited, partially due to their toxicities, although their toxicities are not too high.

MALEIC ANHYDRIDE ($C_4H_2O_3$)

Alternative Names/Abbreviations: 2,5-Furanedione, maleic acid anhydride, toxilic anhydride, cis-butenedioc anhydride, dihydro-2,5-dioxofuran, MA

CAS Registry Number: 108-31-6

Details: Solid (crystalline), combustible, but difficult to ignite. Used in the production of polyester and alkyd resins. With EU Directive 67/548/EEC classification: corrosive (C).

Toxicology: Exposure is by inhalation, MA is a strong irritant (eyes, upper airways and skin), that can produce asthma, and is a proven toxic for animals. MA is not considered as a carcinogen (class A4: not classifiable as a human carcinogen) [39].

TLV-TWA value for MA is low by ACGIH, 2003: 0.1 ppm (corresponding to 0.4 mg/m³) [39-40].

MANGANESE (Mn) AND ITS COMPOUNDS

CAS Registry Number: 7439-96-5

Details: Solid, used in paints and as a preservative in rubber and in wood.

Toxicology: Exposure is by inhalation. It causes neurological disorders in humans, and their chronic exposure can severely effect the central nervous system, having symptoms similar to Parkinson's disease.

Manganese sulfate can cause pancreatic tumours in animals, and there is not enough evidence about the carcinogenicity of manganese and its compounds for humans.

TLV-TWA value for manganese and compounds, as Mn, is low by ACGIH, 2003: 0.2 mg/m³ [41].

MERCURY (Hg) AND COMPOUNDS (ORGANIC AND INORGANIC)

Alternative Names/Abbreviations: Quicksilver, colloidal mercury, liquid silver

CAS Registry Number: 7439-97-6

Details: High density liquids: used as catalyst, as preservatives and in antifouling paints. Mercury is still used in dental amalgams.

Toxicology: Exposure is by inhalation, ingestion and by skin absorption.

Mercury vapour is toxic and it can affect the central nervous system and can cause severe respiratory damage. It accumulates in the kidneys preferentially.

Ingestion of mercuric salts can cause corrosive ulceration and bleeding with problems in the gastrointestinal system, followed by shock and the collapse of circulatory tract.

Its absorption by skin can cause sensitisation dermatitis.

Alkyl compounds of mercury (RHgX) and other organomercury compounds (i.e., methyl mercury) are strong irritants (eye, mucous membranes and skin), the latter with a direct effect on chromosomes. In fact, Minamata disease is termed after the methyl mercury intoxication incidence observed in the fish meat in Minamata, Japan, which produced neurological damage and mental retardation in newborn babies [42].

The International Academy of Oral Medicine and Toxicology (IAOMT) is a Canadian corporation, representing all professionals advocating mercury-free dentistry, the safe removal of mercury dental fillings, and improved standards for dental practices [43].

IARC has announced that there is not enough evidence for the carcinogenicity of mercury and its compounds both in humans and in animals, hence they are A4 - not classifiable as carcinogens [42-44].

TLV-TWA value for Hg and its inorganic compounds are by ACGIH, 2003: 0.025 mg/m^3, as Hg, which is 0.1 mg/m^3 for its aryl compounds.

The corresponding values for organo mercury compounds are TLV-TWA by ACGIH, 2003: 0.01 mg/m^3 as Hg [45, 46].

METHACRYLIC ACID [$C_4H_6O_2$, SMILES CC(C(O)=O)=C]

Alternative Names/Abbreviations: *α-Methacrylic acid, 2-methylene propionic acid, 2-methyl-2-propenoic acid, MAA*

CAS Registry Number: 79-41-4

Details: A corrosive liquid and a monomer (for methacrylic polymers) with an unpleasant acrid odour.

Toxicology: Exposure is by inhalation and by skin absorption. Methacrylic acid is a strong irritant (eyes, upper airways and skin).

No data is available for the carcinogenicity of MAA.

4-METHOXYPHENOL ($C_7H_8O_2$)

Alternative Names/Abbreviations: 4-Hydroxyanisole (in medicine), hydroquinone monomethyl ether

CAS Registry Number: 150-76-5

Details: A solid, an inhibitor (for acrylic monomers), a UV inhibitor and a stabiliser.

Toxicology: Exposure is by inhalation. 4-Methoxy phenol is a cytotoxic and a narcotic at high exposures. It is a carcinogen for animals, but no data is available for humans [47].

TLV-TWA value for 4-methoxy phenol is by ACGIH, 2003: 5 mg/m³.

METHYL ACRYLATE ($CH_2CHCOOCH_3$)

Alternative Names/Abbreviations: Methyl propenoate, acrylic acid methyl ester, 2-propenoic acid methyl ester, MAc

CAS Registry Number: 96-33-3

Details: A liquid and a monomer (for acrylics).

Toxicology: Exposure is by inhalation and by skin absorption. It is a strong irritant (mucous membranes and skin), even at low levels of exposure. The IARC determined that there is insufficient evidence for the carcinogenicity of MAc both to humans and animals [48].

TLV-TWA value for MAc is by ACGIH, 2003: 2 ppm (corresponding to 7 mg/m³).

METHYLACRYLONITRILE [$CH_2C(CH_3)CN$]

Alternative Names/Abbreviations: Methacrylonitrile, iso-propenylnitrile, isopropene cyanide, 2-cyanopropene-1, 2-methyl-2-propenenitrile, MAN

CAS Registry Number: 126-98-7

Details: A liquid and a monomer.

Toxicology: Exposure is by inhalation and by skin absorption. Methylacrylonitrile is a strong neurotoxin [49].

TLV-TWA value for methylacrylonitrile is by ACGIH, 2003: 1 ppm (corresponding to 2.7 mg/m³).

METHYL BUTYL KETONE (C$_6$H$_{12}$O)

Alternative Names/Abbreviations: 2-Hexanone, propyl acetone, n-butyl methyl ketone, MBK, MNBK

CAS Registry Number: 591-78-6

Details: A liquid and a solvent for plastics, with an acetone-like smell (at low concentrations).

Toxicology: Exposure is by inhalation and by skin absorption. An irritant at high concentrations (ocular and respiratory) a toxin for the central nervous system [50].

TLV-TWA value for MBK is by ACGIH, 2003: 5 ppm (corresponding to 20 mg/m^3).

METHYL CHLORIDE (CH$_3$Cl)

Alternative Names/Abbreviations: Chloromethane, monochloromethane

CAS Registry Number: 74-87-3

Details: A gas, a blowing agent for production of cellular plastics.

Toxicology: Exposure is by inhalation. It is a reproductive toxin affecting the central nervous system, causing kidney and liver failures [51]. IARC determined that there is inadequate evidence for its carcinogenic effect to humans, however, NIOSH recommends that methyl chloride be treated as a potential carcinogen [52].

METHYLENE BIS/4-HEXYLISOCYANATE) (C$_{15}$H$_{22}$N$_2$O$_2$)

Alternative Names/Abbreviations: Hydrogenated MDI, dicyclohexylmethane-4, 4-diisocyanate, bis(4-isocyanatocycyclohexyl) methane, HDI

CAS Registry Number: 822-06-0

Details: A liquid and a chemical used in the production of polyurethane polymers.

Toxicology: Exposure is by inhalation. HDI is a strong irritant (eyes, upper airways, nose and skin) [53].

TLV-TWA value for HDI is rather low by ACGIH, 2003: 0.005 ppm (corresponding to 0.054 mg/m^3).

METHYLENE DIPHENYL DIISOCYANATE $CH_2(C_6H_4NO)_2$

Alternative Names/Abbreviations: Diphenylmethane diisocyanate, methylene bisphenyl isocyanate, MDI

CAS Registry Number: 97568-33-7

Details: An aromatic diisocyanate, a liquid and a chemical used in the production of polyurethane polymers. It exists in three isomers (2,2'-MDI, 2,4'-MDI, and 4,4'-MDI). The 4,4' isomer is the most widely used, and is also known as pure MDI. MDI is reacted with a polyol in the manufacture of PU.

Toxicology: Exposure is by inhalation. Although MDI is the least hazardous of the commonly available isocyanates, as it has a very low vapour pressure, it is still a strong irritant (eyes, upper air ways, nose and skin) [54]. IARC determined that there is lack of evidence for the carcinogenicity of MDI to humans. With EU classification, it is harmful (designated Xn).

TLV-TWA value for MDI is also rather low by ACGIH, 2003: 0.005 ppm (corresponding to 0.051 mg/m³).

4,4´-METHYLENE DIANILINE $(C_{13}H_{14}N_2)$

Alternative Names/Abbreviations: 4,4-Amino benzyl aniline, MDA, DDM

CAS Registry Number: 101-77-9

Details: A crystalline solid, and hardening agent used for epoxies.

Toxicology: Exposure is by inhalation, ingestion and skin absorption. MDA is an A3 confirmed carcinogen for animals and a suspected carcinogen for humans. MDA is genotoxic and a human hepato-toxin [55].

TLV-TWA value for MDA is by ACGIH, 2003: 0.1 ppm (corresponding to 0.81 mg/m³).

METHYL ISOAMYL KETONE $[CH_3COCH(C_2H_5)_2]$

Alternative Names/Abbreviations: 5-Methyl-2-hexanone, MIAK

Details: A liquid and a solvent used for polymers.

Toxicology: Exposure is by inhalation. MIAK is an irritant (eyes) and a narcotic (for animals) [56].

TLV-TWA value for MIAK is by ACGIH, 2003: 50 ppm (corresponding to 234 mg/m³).

METHYL ISOBUTYL KETONE ($C_6H_{12}O$)

Alternative Names/Abbreviations: Hexone, 4-methyl, 2-pentanone, MIBK

Details: A Liquid and a solvent used in paints and glues. BIMK has a camphor like odour (detectable at 100 ppm).

Toxicology: Exposure is by inhalation. MIBK is an irritant (eyes, mucous membranes and skin) , and a narcotic, with some mutagenic activity [57].

TLV-TWA value for MIBK is by ACGIH, 2003: 50 ppm (corresponding to 205 mg/m³).

METHYL ISOCYANATE (CH_3CNO)

Alternative Names/Abbreviations: Isocyanic acid methyl ester, MIC

CAS Registry Number: 624-83-9

Details: A liquid and a chemical used for production of PU foams and plastics.

Toxicology: MIC is extremely toxic. Exposure is by inhalation and skin absorption. MIC is an irritant (eyes, mucous membrane and skin) and a toxic at high concentrations that can cause death. It can damage by inhalation, ingestion and contact in concentrations as low as 0.4 ppm. MIC is considered to be a genotoxic that can cause chromosomal abnormalities (as shown by the Bhopal accident in India, when about 43,000 kilograms of MIC were released over a populated area in 1984, killing thousands of people) [58].

TLV-TWA value for MIC is by ACGIH, 2003: 0.02 ppm (corresponding to 0.047 mg/m³).

METHYL METHACRYLATE [$C_5H_8O_2$, SMILES CC(=C)C(=O)OC]

Alternative Names/Abbreviations: Methacrylic acid methyl ester, methyl 2-methylpropenoic acid, MMA

CAS Registry Number: 80-62-6

Details: A liquid and a monomer used for the production of MMA polymers, also used for the production of *co*-polymethyl methacrylate - butadiene - styrene which is used as a modifier for PVC.

Toxicology: Exposure is by inhalation. MMA is a strong irritant (eyes, mucous membrane and skin), and a toxicant affecting nerve cells and causing occupational asthma, foetal toxicities, and possibly cancer [59]. IARC concluded that there is no adequate evidence proving the carcinogenicity of MMA [60].

TLV-TWA value for MMA is by ACGIH, 2003: 50 ppm (corresponding to 205 mg/m³).

2-NAPHTHYLAMINE (C$_{10}$H$_9$N)

Alternative Names/Abbreviations: B-Naphthylamine, 2-aminonaphthalane, BNA

Details: A crystalline solid, used for production of antioxidants and dyes until recently.

Toxicology: Exposure is by inhalation. BNA is a potent carcinogen (bladder). The IARC has concluded that there is enough evidence for the carcinogenicity of BNA both to animals and humans and that it is suggested that exposure by any route be avoided [61].

ACGIH considers BNA as an A1 confirmed carcinogen and no threshold limit value is given.

NICKEL CARBONYL Ni(CO)$_4$

Alternative Name: Nickel tetracarbonyl

CAS Registry Number: 13463-39-3

Details: A liquid and a catalyst used in polymer production.

Toxicology: Exposure is by inhalation. Nickel carbonyl is a strong irritant (pulmonary system).

TLV-TWA value for nickel carbonyl is by ACGIH, 2003: 0.05 ppm (corresponding to 0.12 mg/m^3).

NITROSAMINES (R$_2$N-N=O)

Nitrosamines can be found in a number of rubber and latex products (because of the catalysts and accelerators used), from where it can be released in small amounts over time. Otherwise, nitrosamines are considered to be compounds that form by the reaction of any nitrosating agents (i.e., nitrites) with amines. Hence, in any strongly acidic environment, such as that of the human stomach, nitrosamines can also form. Nitrosamines are found in a number of foodstuffs i.e., in beer, fish, fish by-products, as well as in meat and cheese products, where preservation involves nitrite salts. They can also form when food proteins react with the nitrites in the stomach, as well as by frying the foods or smoking. Although there is a lack of direct evidence for their carcinogenic effects, they are considered to be 'cancer-suspected' agents. The **primary nitrosamines** are unstable and can re-arrange rapidly to form diazoic acids, which have little health threat. Whereas **secondary nitrosamines** can nitrosate rapidly forming persistent nitrosamines, and are the most effective for this. Since a number of traditional cure accelerators of rubber and sulfur donors are products of a number of secondary amines (such as dimethylamine, diethylamine, morpholine or piperidine), they can be released from the accelerator and converted to stable secondary nitrosamines [62, 63].

According to BgVV German legislation, extractable N-nitrosamine/nitrosatables levels should be below 1.0 µg/dm^2, and nitrosamine content should be less than 10 µg/kg [64].

N-NITROSODIPHENYLAMINE ($(C_6H_5)_2N_2O$)

Alternative Name: Diphenyl nitrosamine, NDPhA

CAS Registry Number: 86-30-6

Details: A solid, a vulcanisation retarder/stabiliser used until recently in the rubber industry.

Toxicology: Exposure is by inhalation. NDPhA is an animal carcinogen (as concluded by IARC), and there is not enough evidence for its carcinogenicity to humans [65], with no TLV value established.

There are other stabilisers of a similar nature, which are 4-nitrodiphenylamine, 2-nitrodiphenylamine, N-methyl-*p*-nitroaniline and diphenylamine.

N-NITROSOMORPHOLINE ($C_4H_8N_2O_2$)

Alternative Names/Abbreviations: 4-Nitrosomorpholine, NMOR, NNM

CAS Registry Number: 59-89-2

Details: A crystalline solid, a specific solvent for polyacrylonitrile which is produced during production and processing of rubber.

Toxicology: Exposure is by inhalation and by skin absorption. NMOR is carcinogenic for animals and no evidence is available for its effect on humans.

There is no TLV value established.

PERCHLOROMETHYL MERCAPTAN (CCl_3SCl)

Alternative Names/Abbreviations: Perchloromethane thiol, trichloromethane sulfenyl chloride, PCM

CAS Registry Number: 594-42-3

Details: A liquid and an accelerator used during rubber vulcanisation.

Toxicology: Exposure is by inhalation. PCM is a strong irritant (pulmonary system).

TLV-TWA value for PCM is by ACGIH, 2003: 0.1 ppm (corresponding to 0.76 mg/m^3) [66].

PHENOL (C_6H_5OH, SMILES OC1=CC=CC=C1)

Alternative Names/Abbreviations: Carbolic acid, oxybenzene, phenyl hydroxide, carbolic acid

CAS Registry Number: 108-95-2

Details: A crystalline solid, used in the production of phenolic resins, bisphenol-A and caprolactam.

Toxicology: Exposure is by inhalation, ingestion and through skin absorption. Phenol is an irritant (eyes, mucous membranes and skin), corrosive and toxic to the nervous system as well as to the liver and kidneys. The major hazard of phenol is probably its ability to penetrate the skin rapidly, when liquid, causing severe fatal injuries. Phenol also has a strong corrosive effect on body tissues causing severe chemical burns. Due to its local anaesthetising properties, skin burns can be painless. Phenol is slightly soluble in water, its ingestion acts as a strong poison, with symptoms including a burning pain in mouth and throat, abdominal pain, nausea, vomiting, headache, dizziness, muscular weakness, central nervous system effects, increase in heart rate, irregular breathing, coma, and possibly death. Acute exposure is also associated with kidney and liver damage. Ingestion of one gram has been lethal to humans. Injections of phenol have occasionally been used as a means of rapid execution. IARC concluded that there is inadequate evidence for its carcinogenicity and it cannot be classified as a carcinogen either for animals or for humans [67].

TLV-TWA value for phenol is by ACGIH, 2003: 5 ppm (corresponding to 19 mg/m^3).

N-PHENYL-β-NAPHTYLAMINE ($C_{10}H_7NHC_6H_5$)

Alternative Names: Anilinonaphthalane, PBNA

CAS Registry Number: 135-88-6

Details: A powdery solid, used as an antioxidant in rubber until recently.

Toxicology: Exposure is by inhalation. BPNA is carcinogenic for animals but there is no adequate evidence for such an effect for humans (IARC [68]).

There is no TLV value established for PBNA.

PHTHALIC ANHYDRIDE C$_6$H$_4$(CO)$_2$O

Alternative Names/Abbreviations: Phthalic acid anhydride, phthalandione, 1,2-benzenedicarboxylic acid anhydride, isobenzofuran-1,3-dione

CAS Registry Number: 85-44-9

Details: A crystalline solid, a chemical used for the production of plasticisers.

Toxicology: Exposure is by inhalation. It is an irritant (eyes, skin and upper airways) and can cause asthma. It is not a carcinogen.

TLV-TWA value for phthalic anhydride is by ACGIH, 2003: 1 ppm (corresponding to 6.1 mg/m^3).

POLYBROMINATED BIPHENYLS (C$_{12}$H$_4$Br$_6$/C$_{12}$H$_4$Br$_8$)

Alternative Names/Abbreviations: Brominated biphenyls or polybromobiphenyls (PBB)

Details: Solids, used as flame retarding agents in plastics until 1977, after which it was banned.

Toxicology: Exposure is by inhalation, through skin absorption or ingestion. PBB are proven carcinogenics to animals [69]. Toxicity of PBB towards humans were realised after the accidental farm feed accident in Michigan in 1973 (http://www.michigan.gov/documents/mdch_PBB_FAQ_92051_7.pdf), from which it was understood that PBB can cause some serious dermatological and neurological disorders. Studies in animals showed that PBB can cause weight loss, skin disorders, nervous and severe effects on immune systems, as well as effects on the liver, kidneys, and thyroid gland. IARC accepts that it is carcinogenic to animals.

There are no established TLV values for PBB.

POLYTETRAFLUOROETHYLENE (THERMAL DECOMPOSITION PRODUCTS OF)

Please see polytetrafluorethylene/Teflon in Chapter 9.

PROPYLENE GLYCOL MONOMETHYL ETHER (C$_4$H$_{10}$O$_2$)

Alternative Names/Abbreviations: Propylene glycol methyl ether, PGME

CAS Registry Number: 107-98-2

Details: A liquid with a characteristic bad odour used as a solvent for plastics.

Toxicology: Exposure is by inhalation. PGME is a strong irritant (eyes, upper airways). It can affect the nervous system.

TLV-TWA value for PGME is by ACGIH, 2003: 100 ppm (corresponding to 369 mg/m^3).

PROPYLENE OXIDE (CH$_3$CHOCH$_2$)

Alternative Names/Abbreviations: 1,2-Epoxypropane, propylene epoxide, propene oxide, PO

CAS Registry Number: 75-56-9

Details: A liquid with a characteristic smell of natural gas/ether/benzene, and an epoxide. It is used to produce polyether polyols and the polymer polypropylene oxide (polypropylene glycol) and used as a preservative, and in thermobaric weapons (also called high-impulse thermobaric weapons or fuel-air explosives).

Toxicology: PO is a highly toxic, flammable chemical. Exposure is by inhalation. PO is an irritant (eyes, mucous membranes and skin), and a chemical that causes narcosis. PO is a proven carcinogen for animals [70].

TLV-TWA value for PO is by ACGIH, 2003: 20 ppm (corresponding to 48 mg/m^3).

RADON (Rn)

Please see radon in the Appendix of Chapter 7 (A.7.1)

RUBBER (NATURAL LATEX)

Please see rubber in Chapter 4.

SELENIUM (Se) AND COMPOUNDS

CAS Registry Number: 7782-49-2

Details: Selenium and most of its compounds are solids, with the exception of hydrogen selenide, which is a gas.

Toxicology: Exposure is by inhalation. They are toxic (affecting the central nervous and gastrointestinal systems), with maternal toxicity, and are irritants (eye, upper air ways and skin). High doses are shown to cause tumours (lung and liver) in animals with some mutagenic effects [71].

TLV-TWA value for Se and its compounds is by ACGIH, 2003: 0.2 mg/m³, as Se.

SILICA (CRYSTALLINE)

Alternative Names/Abbreviations: Quartz, tridymite and cristobalit

CAS Registry Number: 112926-00-8

Details: Crystalline silica, or free silica, occurs in nature. It is white or colourless. Silica is the basic component of quartz, sand, granite, and other mineral rocks.

Toxicology: Silica is used in rubber industry as a filler. Inhalation of (crystalline) silica dust is shown to cause 'fibrosis' (scar tissue formation in the lungs) and the disease of 'silicosis' (respiratory disease of the lungs), which may cause disability or death.

IARC concludes that crystalline silica is carcinogenic to humans (Group 1) [72].

STYRENE (MONOMER) ($C_6H_5CHCH_2$, Molecular Formula C_8H_8 , SMILES c1ccccc1C=C)

Alternative Names/Abbreviations: Vinylbenzene, ethenylbenzene, cinnamene, styrol, phenylethylene

CAS Registry Number: 100-42-5

Details: An oily liquid that can evaporate easily and has a sweet smell, it is the monomer for polystyrene and a solvent for synthetic rubber. Its odour threshold is rather low (0.1 ppm).

Toxicology: Exposure is by inhalation, and through skin absorption. Styrene is an irritant (eyes, mucous membranes and skin), a neurotoxic chemical, and a potential carcinogen. IARC concluded that there is limited evidence for the carcinogenicity of styrene both for animals and for humans, with liver damage [73].

TLV-TWA value for styrene is by ACGIH, 2003: 50 ppm (corresponding to 213 mg/m³), with the STEL value of 100 ppm.

STYRENE OXIDE (C_8H_8O)

Alternative Names/Abbreviations: Epoxyethylbenzene, epoxystyrene, phenylethylene oxide, phenyloxirane

CAS Registry Number: 96-09-3

Details: A liquid, used in the production of styrene glycol and in epoxy resins.

Toxicology: Exposure is by inhalation, and through skin absorption. It is an irritant (eyes, and skin) and a maternal toxic. IARC has advised that styrene oxide is a carcinogen for animals and it is a probable carcinogen for humans [74].

There is no TLV value established for styrene oxide.

TETRAHYDROFURAN [$(C_2H_4)_2O$, SMILES C1CCCO1]

Alternative Names/Abbreviations: Tetramethylene oxide, cyclotetramethylene oxide, diethylene oxide, THF

CAS Registry Number: 109-99-9

Details: A heterocyclic organic liquid with an acetone-like smell, and it is a common solvent for plastics.

Toxicology: Exposure is by inhalation. THF is an irritant (upper airways), a central neurotoxic agent and a suspected carcinogen (liver).

TLV-TWA value for THF is by ACGIH, 2003: 200 ppm (corresponding to 590 mg/m^3, with a STEL value of 250 ppm.

TETRAMETHYL SUCCINONITRILE ($C_8H_{12}N_2$)

Alternative Names/Abbreviations: Tetramethylbutanedinitrile, TMSN

CAS Registry Number: 3333-52-6

Details: A crystalline solid, produced from the decomposition of azobis-iso-butrylonitrile, used as blowing agent for cellular plastics production and as a catalyst.

Toxicology: Exposure is by inhalation, and through skin absorption.

TLV-TWA value for TMSN is by ACGIH, 2003: 0.5 ppm (corresponding to 2.8 mg/m^3).

2-MERCAPTOBENZOTHIAZOLE (C_7H_5NS)

Alternative Names/Abbreviations: 1,3-Thiazole, 2-benzothiazolethiol, MBT

CAS Registry Number: 149-30-4

Details: A crystalline solid, and a vulcanisation accelerator of rubber.

Toxicology: Exposure is mainly through skin absorption. MBT is toxic, causing allergic skin reactions (rubber gloves).

No threshold TLV value is established.

4,4′-THIOBIS(6-*tert*-BUTYL-*m*-CRESOL) ($C_{22}H_3O_2S$)
Alternative Names/Abbreviations: Santonox, TBBC
CAS Registry Number: 96-69-5

Details: A powdery solid, used as an antioxidant and stabiliser in plastics packaging materials and in rubber.

Toxicology: Exposure is by inhalation and by ingestion. TBBC has a low toxicity level in animals that can cause contact dermatitis in humans. No carcinogenic effect to humans and animals is observed.

TLV-TWA value for TBBC is by ACGIH, 2003: 10 mg/m³.

THIOGLYCOLIC ACID [$C_2H_4O_2S$, SMILES SCC(=O)O]
Alternative Names/Abbreviations: Mercaptoacetic acid, TGA
CAS Registry Number: 68-11-1

Details: A liquid with a strong unpleasant odour, a stabiliser used in vinylics.

Toxicology: Exposure is by inhalation and by skin absorption. TGA is an irritant and a strong corrosive (eyes, mucous membranes and skin).

TLV-TWA value for TGA is by ACGIH, 2003: 1 ppm (corresponding to 3.8 mg/m³).

THIRAM ($C_6H_{12}N_2S_4$)
Alternative Names/Abbreviations: Tetramethylthiuram disulfide, TMTD
CAS Registry Number: 137-26-8

Details: A crystalline solid, and a vulcanisation accelerator of rubber.

Toxicology: Exposure is by inhalation. TMTD is an irritant (eyes, mucous membranes and skin), causing dermatitis and reproductive failures in animals. Although it is not carcinogenic in some animals, TMTD can react with nitrite and acid (as in the human stomach) to form *N*-nitrosodimethylamine, which is carcinogenic. IARC concludes that there is not enough evidence showing that TMTD is a carcinogen both for animals and humans [75].

TLV-TWA value for thiram is by ACGIH, 2003: 1 mg/m^3.

TIN (Sn) AND ORGANIC COMPOUNDS (ORGANOTIN COMPOUNDS)

Alternative Names/Abbreviations: Triethyltin iodide, dibutyltin chloride, triphenyltin acetate, bis(tributyltin) oxide, triphenyltin chloride

Details: Silvery, malleable, poor metal and an element. Compounds of tin are solids or liquids. They are mostly used as stabilisers and as biocides in polymers.

Toxicology: Exposure is by inhalation and by skin absorption. Certain organic tin compounds (organotins) such as triorganotins are toxic and are used as industrial fungicides and bactericides. All organotin compounds are irritant (eyes, mucous membranes and skin), and toxic. The most toxic ones are trialkyltin compounds, such as tributyltin oxide, followed by dialkyl and monoalkyltin compounds. In each group, the ethyl derivatives are the most toxic ones. They have maternal toxicities to different extent [76].

TLV-TWA value for organotin compounds is by ACGIH, 2003: 0.1 mg/m^3), as Sn, and with a STEL value of 0.2 mg/m^3.

TITANIUM DIOXIDE (TiO$_2$, SMILES O=Ti=O)

Alternative Names/Abbreviations: Rutile, anatase, octahedrite, unitane, titanium (IV) oxide, titania, titanium white, pigment white 6 (CI 77891)

CAS Registry Number: 13463-67-7

Details: A white fine powdery solid, used in plastics and paints as a pigment of high refringence. It acts as a UV reflector, hence it can provide UV resistance. It is also a photocatalyst.

Toxicology: Exposure is by inhalation. Rutile is a mild irritant (pulmonary system), there are no incidences observed for its carcinogenicity, mutagenicity or genotoxicity.

TLV-TWA value for titanium dioxide is by ACGIH, 2003: 10 mg/m^3.

TOLUENE 2,4-DIISOCYANATE $CH_3C_6H_3(NCO)_2$

Alternative Names/Abbreviations: Toluene diisocyanate, two isomers: 2,4-TDI and 2,6-TDI; 2,4-diisocyanato-1-methyl-benzene, TDI

CAS Registry Number: 584-84-9

Details: A liquid, used in the production of PU plastics and foams.

Toxicology: Exposure is by inhalation. TDI is a strong irritant (eyes, mucous membranes and skin) and a strong sensitiser for the upper airways. Exposure of humans to TDI first causes eye, nose and throat irritation, followed by chest pain with respiratory symptoms, bronchitis, nausea and vomiting (at a concentration around 0.5 ppm). TDI can induce asthma, and has a carcinogenic effect (pancreatic) in animals. The IARC have decided that it is carcinogenic to animals, but there is not enough evidence for its carcinogenicity to humans [77].

TLV-TWA value for TDI is by ACGIH, 2003: 0.005 ppm (corresponding to 0.036 mg/m³), with a STEL value of 0.02 ppm.

TRIBUTYL PHOSPHATE $(C_4H_9)_3PO_4$

Alternative Names/Abbreviations: n-Tributyl phosphate or tri-n-butyl phosphate, phosphoric acid tributyl ester, TBP

CAS Registry Number: 126-73-8 (anhydrous), and 6131-90-4 (trihydrate)

Details: An odourless liquid, used as a plasticiser (particularly for cellulose esters, i.e., nitrocellulose and cellulose acetate), and as an antifoaming agent.

Toxicology: Exposure is by inhalation. TBP is a strong irritant (eyes, mucous membranes and skin). Inhalation and ingestion can have effects on the central nervous system.

TLV-TWA value for TBP is by ACGIH, 2003: 0.2 ppm (corresponding to 2.2 mg/m³).

TRIETHYLENE TETRAMINE $(C_6H_{18}N_4)$

Alternative Names/Abbreviations: Araldite hardener, TETA

CAS Registry Number: 112-24-3

Details: A viscous oily liquid, used as a hardener and crosslinker for epoxy resin curing.

Toxicology: Exposure is by inhalation. TETA is a strong irritant (eyes, mucous membranes, skin) and a sensitiser for the upper airways, causing asthma. It is thought to be mutagenic to animals.

No TLV value is established for TETA.

TRIMELLITIC ANHYDRIDE ($C_9H_4O_5$)

Alternative Names/Abbreviations: Anhydromellitic acid, TMA, TMAN

CAS Registry Number: 552-30-7

Details: A crystalline solid, used as a curing agent for epoxies, and as a plasticiser for vinylics.

Toxicology: Exposure is by inhalation. TMA is an irritant (respiratory system). Irritation is usually followed by asthma, and at the later stage by fever and aching muscles and joints (TMA flu), and anaemia. There are no reports for its carcinogenicity or mutagenicity.

TLV-Ceiling value for TMA is by ACGIH, 2003: 0.04 mg/m³.

TRIMETHYL BENZENE ($(CH_3)_3C_6H_3$)

Three isomers of trimethyl benzene (TMB) are *mesitylene* 1,3,5-TMB (CAS Registry Number: 108-67-8), *pseudocumene* 1,2,4-TMB (CAS Registry Number: 95-63-6) and *hemimellitene* 1,2,3-TMB (CAS Registry Number: 526-73-8).

Details: A liquid, used as a UV stabiliser in plastics.

Toxicology: Exposure is by inhalation. TMB is an irritant (eyes, nose and respiratory system), and a central nervous system depressant.

TLV-TWA value for TMB is by ACGIH, 2003: 25 ppm (corresponding to 123 mg/m³).

TRIMETHYL PHOSPHITE ($C_3H_9O_3P$)

Alternative Names/Abbreviations: Methyl phosphite, trimethoxy phosphine, TMP

Details: A liquid, used as a flame retardant in plastics.

Toxicology: Exposure is by inhalation and through skin absorption. TMP is an irritant (skin) and can produce ocular damage. It is genotoxic to some animals [78].

TLV-TWA value for TMP is by ACGIH, 2003: 2 ppm (corresponding to 10 mg/m³).

TRIORTHOCRESYL PHOSPHATE $(CH_3C_6H_4)_3PO_4$

Alternative Names/Abbreviations: Tri-o-tolyl phosphate, TOCP

CAS Registry Number: 78-30-8

Details: A liquid, used as plasticiser in vinylics and as a flame retardant.

Toxicology: Exposure is by inhalation, ingestion and through skin absorption. TOCP is shown to cause flaccid and spastic paralysis (after ingestion of 6-7 mg/kg [79]), and is toxic. It causes nausea first followed by vomiting, diarrhoea and abdominal pain. Its effect is usually from 'accidental ingestion' of contaminated foods and beverages (such as the Jamaican ginger extract 'Jake' [80]), but reports of poisoning from exposure to it is almost nil.

TLV-TWA value for TOCP is by ACGIH, 2003: 0.1 mg/m^3.

TRIPHENYL PHOSPHITE $P(O-C_6H_5)_3$

Alternative Names/Abbreviations: Phosphorus acid triphenyl ester, TPP

CAS Registry Number: 101-02-0

Details: A solid (below 22 °C) or a liquid, used as a stabiliser and antioxidant (for vinylics and rubber).

Toxicology: Exposure is by inhalation. TPP is an irritant (eye and skin) for humans and it is a neurotoxic for animals.

TLV value of TPP is not established.

n-VALERALDEHYDE $(C_5H_{10}O)$

Alternative Names/Abbreviations: Amyl aldehyde, pentanal, valeric aldehyde

CAS Registry Number: 110-62-3

Details: A liquid with a specific odour at 0.028 ppm threshold level. It is used as accelerator for rubber vulcanisation.

Toxicology: Exposure is by inhalation and ingestion. It is an irritant (eyes and skin), and is toxic, causing chromosomal and DNA effects in some animals, but no effects to humans is reported.

TLV-TWA value for *n*-valeraldehyde is by ACGIH, 2003: 50 ppm (corresponding to 176 mg/m^3).

VINYL ACETATE (C$_4$H$_6$O$_2$, SMILES C=COC(C)=O)

Alternative Names/Abbreviations: Vinyl acetate monomer, vinyl ethanoate, 1-acetoxyethylene, VA

CAS Registry Number: 108-05-4

Details: A liquid and a monomer that are used for the production of polyvinyl acetate (PVAc) polymers and copolymers (such as ethylene-vinyl acetate).

Toxicology: Exposure is by inhalation. Vinyl acetate is an irritant (eyes, nose, throat and skin), a suspected carcinogen and genotoxic for animals. IARC concluded that there is limited evidence for animals and inadequate evidence for humans for the carcinogenicity of vinyl acetate [81].

TLV-TWA value for vinyl acetate is by ACGIH, 2003: 10 ppm (corresponding to 35 mg/m^3), with a STEL value of 15 ppm.

VINYL CHLORIDE (C$_2$H$_3$Cl)

Alternative Names/Abbreviations: Vinyl chloride monomer, chloroethene, chloroethylene, ethylene monochloride, VCM

CAS Registry Number: 75-01-4

Details: A colourless gas with a sweet odour and a monomer used for the production of polyvinyl chloride PVC polymers.

Toxicology: Exposure is by inhalation. Inhaling its vapours produces symptoms similar to alcohol intoxication. VCM is a carcinogen (a rare tumour, liver angiosarcoma), a genotoxic, and a central nervous system depressant. Chronic exposure which occurs at high levels of VCM vapour, can cause 'Vinyl Chloride Disease' [82]. The tumourigenic effect of VCM has already been confirmed many times for animals. Exposure to vinyl chloride during pregnancy in animals has produced miscarriages and birth defects, however, its effect on human reproduction is not known. VCM is considered to be a carcinogen specifically linked to certain cancers of the liver (angiosarcoma and hepatocellular carcinoma). IARC has determined that there is sufficient evidence that VCM is carcinogenic both to animals and to humans, and it has been given an A1 rating certain carcinogen for humans [83].

TLV-TWA value for VCM is by ACGIH, 2003: 1 ppm (corresponding to 3 mg/m^3).

VINYL BROMIDE (C₂H₃Br)

Alternative Names/Abbreviations: Bromoethene, bromoethylene, ethylene monobromide

CAS Registry Names: 593-60-2

Details: A gas and a monomer used for the production of polyvinyl bromide polymers.

Toxicology: Exposure is by inhalation. Vinyl bromide is carcinogenic, mutagenic, and a central nervous system depressant for animals. No complete data is available for its effects on humans, and the IARC concluded that it is a carcinogen for animals and an A2-suspected probable carcinogen for humans [84].

TLV-TWA value for vinyl bromide is by ACGIH, 2003: 0.5 ppm (corresponding to 2.2 mg/m³).

4-VINYLCYCLOHEXENE (C₈H₁₂)

Alternative Names/Abbreviations: 4-Ethenylcyclohexene, VCH

CAS Registry Number: 100-40-3

Details: A liquid and an intermediate found in the production of flame retardants and during cure of rubber in tyre processing.

Toxicology: Exposure is by inhalation and through skin absorption. VCH is an irritant (moderately for skin) and a carcinogen for some animals. IARC announced that VCH is a carcinogen for animals, and an A2-suspected carcinogen for humans [85].

TLV-TWA value for VCH is rather low by ACGIH, 2003: 0.1 ppm (corresponding to 0.4 mg/m³).

VINYL CYCLOHEXENE DIOXIDE (C₈H₁₂O₂)

Alternative Names/Abbreviations: 4-Vinylcyclohexene diepoxide, VCD

Details: A liquid and a diluent for epoxy resins.

Toxicology: Exposure is by inhalation and through skin absorption. VCD is an irritant (eyes, upper air ways and skin) and carcinogenic to animals. IARC announced that VCH is a carcinogen for animals, and a B2-suspected possible carcinogen for humans. [86].

TLV-TWA value for VCD is by ACGIH, 2003: 10 ppm (corresponding to 57 mg/m³).

VINYLIDENE CHLORIDE [$C_2H_2Cl_2$, SMILES ClC(Cl)=C]

Alternative Names/Abbreviations: Vinylidene dichloride, 1,1-dichloroethylene (1,1-DCE), asymmetric-dichloroethylene, VDC

CAS Registry Number: 75-35-4

Details: A highly flammable liquid, used for production of polyvinylidene chloride polymers and co-polymers.

Toxicology: Exposure is by inhalation. VDC is an irritant (upper airways, eyes, skin), a genotoxic, and a central nervous system depressant (with symptoms of sedation, inebriation, convulsions, spasms, and unconsciousness at high concentrations that can cause damage to kidney and liver in animals). There is limited data available for its effect on humans.

TLV-TWA value for VDC is by ACGIH, 2003: 5 ppm (corresponding to 20 mg/m^3), with a STEL value of 20 ppm.

ZINC DITHIOCARBAMATES

Alternative Names/Abbreviations: Zinc-diethylthiocarbamate (ZDEC) dibutylthiocarbamate (ZDBC), dimethyldithiocarbamate (ZDMC), ZDC

Details: Rapid vulcanisation accelerators for rubbers.

Toxicology: Exposure is through skin absorption. ZDC are skin-contact allergens that can cause type IV delayed type hypersensitivities, and are mutagenic.

There is no available TLV data established for ZDC.

References

1. L. Silverman, H.F. Schulte and M.W. First, *Journal of Industrial Hygiene and Toxicology*, 1946, **28**, 262.

2. *Chemical Hazard Information Profile: Acetaldehyde*, Environmental Protection Agency, Washington, DC, USA, 1983.

3. *Re-evaluation of Some Organic Chemicals, Hydrazine and Hydrogen Peroxide*, IARC Monographs on the Evaluation of the Carcinogenic Risk of Chemicals to Humans, Volume 71, IARC, Lyon, France, 1999, p.1211.

4. J.S. Sinclair, D.T. McManus, M.D. O'Hara and R. Millar, *Burns: Journal of the International Society for Burn Injuries*, 1994, **20**, 5, 469.

5.   *Criteria for a Recommended Standard: Occupational Exposure to Nitriles*, Publication No.78-212, DHHS (NIOSH), Washington, DC, USA, 1978.

6.   G.J. Hathaway and N.H. Proctor, *Proctor and Hughes Chemical Hazards of the Workplace, 5th Edition*, Wiley Interscience, Hoboken, NJ, USA, 2004, p.19.

7.   News and Alerts, Office of the Attorney General, Bill Lockyer, Department of Justice, http://ag.ca.gov/newsalerts/release.php?id=1207

8.   A.M. Prentiss, *Chemicals in War: a Treatise of Chemical Warfare*, McGraw Hill, New York, NY, USA, 1937, p.139.

9.   C.C. Willhite, *Journal of Applied Toxicology*, 1982, **2**, 1, 54.

10. T.R. Torkelson and V.K. Rowe in *Patty's Industrial Hygiene and Toxicology*, Eds., G.D. Clayton and F.R. Clayton, Wiley Interscience, New York, NY, USA, 1981, p.3568.

11. US Department of Labor: Occupational Safety and Health Standards-Carcinogens, Federal Regulations, http://www.gpoaccess.gov/cfr/index.html

12. *Documentation of the Threshold Limit Values and Biological Exposure Indices*, 7th Edition, ACGIH Publications, Cincinnati, OH, USA, 2001.

13. W. Bi, R.B. Hayes and P. Feng, *American Journal of Industrial Medicine*, 1992, **21**, 4, 481.

14. *Beryllium, Cadmium, and Mercury and Exposures in the Glass Manufacturing Industry*, IARC Monographs on the Evaluation of Carcinogenic Risks to Humans, Volume 58, IARC, Lyon, France, 1993, p.119.

15. *Printing Processes and Printing Inks, Carbon Black and some Nitrocompounds*, IARC Monographs on the Evaluation of Carcinogenic Risks to Humans, Volume 65, IARC, Lyon, France, 1996, p.149.

16. *Re-evaluation of some Organic Chemicals Hydrazine and Hydrogen Peroxide*, IARC Monographs on the Evaluation of Carcinogenic Risks to Humans, Volume 71, IARC, Lyon, France, 1999, p.433.

17. *Polychlorinated Dibenzo-para-Dioxins and Polychlorinated Dibenzofurans*, IARC Monographs on the Evaluation of Carcinogenic Risks to Humans, Volume 69, IARC, Lyon, France, 1997, p.33.

18. *Re-evaluation of some Organic Chemicals Hydrazine and Hydrogen Peroxide*, IARC Monographs on the Evaluation of Carcinogenic Risks to Humans, Volume 71, IARC, Lyon, France, 1999, p.227.

19. *Chromium, Nickel and Welding*, IARC Monographs on the Evaluation of Carcinogenic Risks to Humans, Volume 49, IARC, Lyon, France, 1990, p.49.

20. *Chlorinated Drinking Water; Chlorination By-products; Some Other Halogenated Compounds; Cobalt and Cobalt Compounds*, IARC Monographs on the Evaluation of Carcinogenic Risks to Humans, Volume 52, IARC, Lyon, France, 1991, p.363.

21. *Toxicological Profile for Cresols*, ATSDR Monographs, US Department of Health and Human Services, Public Health Service, TP/91-11, 1992.

22. D.S. Marsman, *NTP Technical Report on Toxicity Studies of Dibutyl Phthalate administered in Feed to F344 Rats and B6C3F₁ Mice*, NIH Publication no.95-3353, NIH, Bethesda, MD, USA, 1995.

23. G.J. Hatcaway and N.H. Proctor, *Proctor and Hughes Chemical Hazards of the Workplace*, 5th Edition, Wiley Interscience, Hoboken, NJ, USA, 2004.

24. *Some Industrial Chemicals*, IARC Monographs on the Evaluation of Carcinogenic Risks to Humans, Volume 77, IARC, Lyon, France, 2000, p.41.

25. *Toxicological Profile for DEHP*, ATSDR Monographs, US Department of Health and Human Services, Public Health Service, 2002, http://www.atsdr.cdc.gov/toxprofiles/tp9.html

26. *Occupational Safety and Health Guideline for DGE as Potential Human Carcinogen,* Occupational Safety and Health Guidelines for Chemical Hazards, Supplement II, OHG. Publication No.89-104.1-6, NIOSH, Cincinnati, OH, USA, 1988.

27. *Occupational Exposures of Hairdressers and Barbers and Personal Use of Hair Colorants; Some Hair Dyes; Cosmetic Colorants; Industrial Dyestuffs and Aromatic Amines,* IARC Monographs on the Evaluation of Carcinogenic Risks to Humans, Volume 57, IARC, Lyon, France, 1993, p.357.

28. *Re-evaluation of some Organic Chemicals Hydrazine and Hydrogen Peroxide,* IARC Monographs on the Evaluation of Carcinogenic Risks to Humans, Volume 71, IARC, Lyon, France, 1999, p.1437.

29. *Isocyanates*, NIOSH, Washington, DC, USA, http://www.cdc.gov/niosh/topics/isocyanates/

30. *Isocyanates*, Occupational Safety and Health Administration, http://www.osha.gov/dts/chemicalsampling/data/CH_248525.html

31. *Re-evaluation of some Organic Chemicals Hydrazine and Hydrogen Peroxide*, IARC Monographs on the Evaluation of Carcinogenic Risks to Humans, Volume 71, IARC, Lyon, France, 1999, p.589.

32. *Re-evaluation of some Organic Chemicals Hydrazine and Hydrogen Peroxide*, IARC Monographs on the Evaluation of Carcinogenic Risks to Humans, Volume 71, IARC, Lyon, France, 1999, p.603.

33. *Some Thyrotrophic Agents*, IARC Monographs on the Evaluation of Carcinogenic Risks to Humans, Volume 79, IARC, Lyon, France, 2001, p.659.

34. *Wood Dust and Formaldehyde*, IARC Monographs on the Evaluation of Carcinogenic Risks to Humans, Volume 62, IARC, Lyon, France, 1995, p.217.

35. *Toxicological Profile for HDI 157*, ASTDR Registry, US Department of Health and Human Services, Public Health Service, 1998.

36. *Re-evaluation of some Organic Chemicals, Hydrazine and Hydrogen Peroxide*, IARC Monographs on the Evaluation of Carcinogenic Risks to Humans, Volume 71, IARC, Lyon, France, 1999, p.1465.

37. *Report on Carcinogens*, Eleventh Edition, US Department of Health and Human Services, Public Health Service, National Toxicology Program, USA, 2005.

38. I. Lancranjan, H.I. Popescu, O. Gavanescu, I. Klepsch and M. Sebanescu, *Archives of Environmental Health*, 1975, **30**, 8, 396.

39. *Maleic Anhydride*, International Occupational Safety and Health Information Centre, International Labour Organisation, www.ilo.org/public/english/protection/safework/cis/products/icsc/dtasht/_icsc07/icsc0799.htm

40. *Chronic Dietary Administration of Maleic Anhydride*, CIIT Final Report Docket No.114N3, CIIT Centers for Health, Research Triangle Park, NC, USA, 1983.

41. *Toxicological Profile for Manganese*, ATSDR Report, US Department of Health and Human Services, Public Health Service, 2000.

42. *Minamata Disease: The History and Measures*, www.env.go.jp/en/chemi/hs/minamata2002/refer.html

43. *International Academy of Oral Medicine and Toxicology*, www.iaomt.org

44. *Beryllium, Cadmium, Mercury and Exposures in the Glass Manufacturing Industry*, IARC Monographs on the Evaluation of Carcinogenic Risks to Humans, Volume 58, IARC, Lyon, France, 1993, p.289.

45. *Beryllium, Cadmium, Mercury and Exposures in the Glass Manufacturing*, Industry, IARC Monographs on the Evaluation of Carcinogenic Risks to Humans, Volume 58, IARC, Lyon, France, 1993, p.289.

46. *Report on the Toxicological Profile for Mercury*, ATSD Registry, US Department of Health and Human Services, Public Health Service, USA, 1999.

47. E. Asakawa, M. Hirose, A. Hagiwara, S. Takahashi and N. Ito, *International Journal of Cancer*, 1994, **56**, 1, 146.

48. *Re-evaluation of some Organic Chemicals Hydrazine and Hydrogen Peroxide*, IARC Monographs on the Evaluation of Carcinogenic Risks to Humans, Volume 71, Lyon, France, 1999, p.1489.

49. *Toxicology and Carcinogenesis Studies of Methacrylonitrile F344 Rats and B6C3F$_1$ Mice (Gavage Studies)*, NTP Technical Report Series, No.497, 2001, p.1-226.

50. *Toxicological Profile for 2-Hexanone*, ATSDR Report, US Department of Health and Human Services, Public Health Service, 1992.

51. *Re-evaluation of some Organic Chemicals Hydrazine and Hydrogen Peroxide*, IARC Monographs on the Evaluation of Carcinogenic Risks to Humans, Volume 71, IARC, Lyon, France, 1999, p.737.

52. *Monohalomethanes, Methyl Chloride-Methyl Bromide-Methyl Iodide*, NIOSH, Current Intelligence Bulletin 43, Department of Health and Human Services, 1984.

53. W.L. Richards, *Generic Health Hazard Assessment of the Chemical Class Diisocyanates*, EPA Contract No.68-02 3990, EPA, Washington, DC, USA, 1987.

54. *Re-evaluation of some Organic Chemicals, Hydrazine and Hydrogen Peroxide*, IARC Monographs on the Evaluation of Carcinogenic Risks to Humans, Volume 71, IARC, Lyon, France, 1999, p.1049.

55. *Some Chemicals Used in Plastics and Elastomers*, IARC Monographs on the Evaluation of Carcinogenic Risks to Humans, Volume 39, IARC, Lyon, France, 1985 p.349.

56. W.J. Krasavage, J.L. Odonoghua and G.D. Divicenzo in *Patty's Hygiene and Toxicology*, Eds., F.E. Clayton and G.D. Clayton, Wiley Interscience, New York, NY, USA, 1982.

57. *American Industrial Hygiene Association Journal*, 1966, **27**, 209.

58. *Methyl Isocyanate Documentation of TLVs and BEIs*, ACGIH, Cincinnati, USA, 1991, p.1022.

59. H.M. Solomon, J.E. McLaughlin, R.E. Swenson, V. Hagan, F.J. Wanner, G.P. O'Hara and N.D. Krivanek, *Teratology*, 1993, **48**, 2, 115.

60. *Some Industrial Chemicals*, IARC Monographs on the Evaluation of Carcinogenic Risks to Humans, Volume 60, IARC, Lyon, France, 1994, p.445.

61. *Overall Evaluations of Carcinogenicity: An Updating of IARC Monographs Volumes 1-42*, Supplement 7, IARC, Lyon, France, 1987, p.261.

62. D.W. Chasar, *Rubber and Plastics News,* 1994, **23**, 17, 15.

63. J.R. Loadman in *Proceedings of the IRC 93/144th Meeting*, Orlando, FL, USA, 1993, Paper No.16.

64. J.A. Sidwell and M.J. Forrest, *Rubbers in Contact with Food*, Rapra Review Report No.119, 2000, Volume 10, No.11.

65. *Some Aromatic Amines; Anthroquinones and Nitroso Compounds and Inorganic Fluorides Used in Drinking Water and Dental Preparations*, IARC Monographs on the Evaluation of Carcinogenic Risks to Humans, Volume 27, IARC, Lyon, France, 1982, p.213.

66. *Perchloro Mercaptan Documentation of the TLV's and BEI's*, ACGIH, Cincinnati, USA, 1991, p.1195.

67. *Re-evaluation of some Organic Chemicals Hydrazine and Hydrogen Peroxide,* IARC Monographs on the Evaluation of Carcinogenic Risks to Humans, Volume 71, IARC, Lyon, France, 1999, p.749.

68. *Some Aromatic Amines and Related Nitro Compounds – Hair dyes, Colouring Agents and Miscellaneous Industrial Chemicals,* IARC Monographs on the Evaluation of Carcinogenic Risks to Humans, Volume 16, IARC, Lyon, France, 1978, p.325.

69. *Some Halogenated Hydrocarbons and Pesticide Exposures*, IARC Monographs on the Evaluation of Carcinogenic Risks to Humans, Volume 41, IARC, Lyon, France, 1986, p.262.

70. *Some Industrial Chemicals*, IARC Monographs on the Evaluation of Carcinogenic Risks to Humans, Volume 60, IARC, Lyon, France, 1994, p.181.

71. *Toxicological Profile for Selenium*, Agency for Toxic Substances and Disease Registry, US Department of Health and Human Services Public Health Service, Atlanta, GA, USA, 2001.

72. *Silica, Some Silicates, Coal Dust and Para-Aramid Fibrils*, IARC Monographs on the Evaluation of Carcinogenic Risks to Humans, Volume 68, IARC, Lyon, France, 1997.

73. *Some Industrial Herbal Medicines, Some Mycotoxins, Naphthalene and Styrene*, IARC Monographs on the Evaluation of Carcinogenic Risks to Humans, Volume 82, IARC, Lyon, France, 2002, p.437.

74. *Some Industrial Chemicals*, IARC Monographs on the Evaluation of Carcinogenic Risks to Humans, Volume 60, IARC, Lyon, France, 1994, p.321.

75. *Occupational Exposures in Insecticide Application and Some Pesticides*, IARC Monographs on the Evaluation of Carcinogenic Risks to Humans, Volume 53, IARC, Lyon, France, 1991, p.403.

76. *Toxicological Profile for Tin*, ASTDR, US Department of Health and Human Services, Public Health Service, 1992.

77. *Re-evaluation of some Organic Chemicals Hydrazine and Hydrogen Peroxide*, IARC Monographs on the Evaluation of Carcinogenic Risks to Humans, Volume 71, IARC, Lyon, France, 1999, p.865.

78. *Trimethyl Phosphite-Documentation of the TLVs and BEIs*, ACGIH, Cincinnati, OH, USA, 1991, p.1650.

79. D.W. Fassett in *Industrial Hygiene and Toxicology*, Ed., F.A. Patty, Wiley Interscience, New York, NY, USA, 1963, p.1853.

80. J.P. Morgan and T.C. Tulloss, *Annals of Internal Medicine*, 1976, 85, 804.

81. *Dry Cleaning, Some Chlorinated Solvents and Other Industrial Chemicals*, IARC Monographs on the Evaluation of Carcinogenic Risks to Humans, Volume 63, IARC, Lyon, France, 1995, p.443.

82. Agency for Toxic Substances and Disease Registry, *Toxicological Profile for Vinyl Chloride*, ATSDR Monographs, US Department of Health and Human Services Public Health Service, USA, 2006.

83. *Overall Evaluations of Carcinogenicity: An Updating of IARC Monographs* Volume 1-42, Supplement 7, IARC, Lyon, France, 1987, p.373.

84. *Re-evaluation of some Organic Chemicals Hydrazine and Hydrogen Peroxide,* IARC Monographs on the Evaluation of Carcinogenic Risks to Humans, Volume 71, IARC, Lyon, France, 1999, p.923.

85. *Some Industrial Chemicals,* IARC Monographs on the Evaluation of Carcinogenic Risks to Humans, Volume 60, IARC, Lyon, France, 1994, p.357.

86. *Some Industrial Chemicals,* IARC Monographs on the Evaluation of Carcinogenic Risks to Humans, Volume 60, IARC, Lyon, France, 1994, p.347.

Some Additional Related References

1. *Sigel's Series on Metal Ions in Life Sciences, Eds.,* A. Sigel, H. Sigel and R. Sigel, John Wiley and Sons Ltd, Sussex, UK, 2007.

2. Substance Fact Sheet, The National Pollutant Inventory of Australia, http://www.npi.gov.au/database/substance-info/profiles/6.html

3. Chemical Search Engine, PubChem, http://www.chemindustry.com/chemicals.html

4. *The Polyurethanes Book,* Eds., D. Randall and S. Lee, Wiley, New York, NY, USA, 2002.

11 Short Lists of Some Extremely Hazardous Substances and IARC Groups 1, 2a, 2b, 3 and 4 Carcinogens Related to Plastics and Rubbers

11.1 A List of Some Extremely Hazardous Substances Related to Plastics and Rubbers

'Extremely Hazardous Substances' are as defined by Section 302 of the US Emergency Planning and Community Right-to-Know Act.
http://yosemite.epa.gov/oswer/ceppoehs.nsf/EHS_Profile?openform

Acrolein	Hydroquinone
Acrylamide	Mercuric acetate
Acrylonitrile	Mercuric chloride
Cadmium stearate	Methacrolein diacetate
Chloroform	Methacrylic anhydride
Diepoxybutane	Methacrylonitrile
Diglycidyl ether	Methyl isocyanate
Dimethyl-*p*-phenylenediamine	Methyl vinyl ketone
Dimethyldichlorosilane	Nitrosodimethylamine
Epichlorohydrin	Phenol
Ethylene oxide	Tetraethyllead
Ethylenediamine	Tetraethyltin
Formaldehyde	Toluene 2,4-diisocyanate
Formaldehyde cyanohydrin	Toluene 2,6-diisocyanate
Furan	Trimethyltin chloride
Hexachlorocyclopentadiene	Triphenyltin chloride
Hexamethylenediamine, *n,n´*-dibutyl-	Tris(2-chloroethyl) amine
Hydrocyanic acid	Vinyl acetate monomer
Hydrogen sulfide	

11.2 A Brief List of IARC Group 1 Carcinogens for Chemicals Related to Plastics and Rubbers

Based on *IARC Monographs* (http://monographs.iarc.fr/index.php)

IARC *Group 1: means the agent is carcinogenic to humans. The exposure circumstance entails exposures that are carcinogenic to humans.*

Arsenic and arsenic compounds	Ethylene oxide
Asbestos	Formaldehyde
Benzene	Radon-222 and its decay products
Beryllium and its compounds	2,3,7,8-Tetrachlorodibenzo-*para*-dioxin
Cadmium and its compounds	Vinyl chloride (VCM)
Dioxin	

11.3 A Brief List of IARC Group 2A Carcinogens for Chemicals Related to Plastics and Rubbers

Based on *IARC Monographs* (http://monographs.iarc.fr/index.php)

IARC *Group 2A: means the agent is probably carcinogenic to humans. The exposure circumstance entails exposures that are probably carcinogenic to humans.*

Acrylamide	Tetrachloroethylene
1,3-Butadiene	Trichloroethylene
Epichlorohydrin	Vinyl bromide
Ethylene dibromide	Vinyl fluoride
Lead compounds	
Inorganic polychlorinated biphenyls	

11.4 A Brief List of IARC Group 2B Carcinogens for Chemicals

Related to Plastics and Rubbers

Based on *IARC Monographs* (http://monographs.iarc.fr/index.php)

IARC *Group 2B: means the agent is possibly carcinogenic to humans. The exposure circumstance entails exposures that are possibly carcinogenic to humans.*

Acetaldehyde

Acetamide

Acrylonitrile

Antimony trioxide

Carbon black

Carbon tetrachloride

Chloroform

Chloroprene

Dichloromethane (methylene chloride)

1,4-Dioxane

1,2-Epoxybutane

Ethyl acrylate

Ethylbenzene

Isoprene

Lead

Methylmercury compounds

Phenyl glycidyl ether

Propylene oxide

Styrene

Tetrafluoroethylene

Titanium dioxide

Toluene diisocyanates

Urethane

Vanadium pentoxide

Vinyl acetate

11.5 A Brief List of IARC Group 3 Carcinogens for Chemicals Related to Plastics and Rubbers

Based on *IARC Monographs* (http://monographs.iarc.fr/index.php)

IARC Group 3: means the agent is not classifiable as to its carcinogenicity to humans. This category is used most commonly for agents and exposure circumstances for which the evidence of carcinogenicity is inadequate in humans and inadequate or limited in experimental animals.

Acrolein

Acrylic acid

Acrylic fibres

Acrylonitrile-butadiene-styrene copolymers

Allyl chloride

Allyl isothiocyanate

Antimony trisulfide

Para-aAramid fibrils

Benzoyl peroxide

Bisphenol-A diglycidyl ether

n-Butyl acrylate

Nylon 6 (PA6)

Phenol

Polyacrylic acid

Polychlorinated dibenzo-*para*-dioxins (other than 2,3,7,8-tetrachlorodibenzo-*para*-dioxin)

Polychloroprene

Polyethylene (PE)

Polymethylene polyphenyl isocyanate

Polymethyl methacrylate (PMMA)

Polypropylene (PP)

Polystyrene (PS)

Dichloroacetylene

meta/ortho-Dichlorobenzene

Di(2-ethylhexyl) adipate (DEHA)

Di(2-Ethylhexyl) phthalate (DEHP)

2-Ethylhexyl acrylate

Hydroquinone

Insulation glass wool

Lauroyl peroxide

Lead compounds, organic

Melamine

Mercury and inorganic mercury compounds

Methyl acrylate, (MA)

Methyl methacrylate, (MMA)

N-Methylolacrylamide

Modacrylic fibres

1,5-Naphthalene diisocyanate

Polytetrafluoroethylene (Teflon)

Polyurethane foams

Polyvinyl acetate

Polyvinyl alcohol

Polyvinyl chloride (PVC)

Polyvinyl pyrrolidone

Selenium and selenium compounds

Styrene-acrylonitrile copolymers (SAN)

Styrene-butadiene copolymers (SBR)

Titanium dioxide

Vinyl chloride - vinyl acetate copolymers

Vinylidene chloride

Vinylidene chloride-vinyl chloride copolymers

Vinylidene fluoride

N-Vinyl-2-pyrrolidone

Vinyl toluene

11.6 A Brief List of IARC Group 4 Carcinogens for Chemicals Related to Plastics and Rubbers

Based on *IARC Monographs* (http://monographs.iarc.fr/index.php)

IARC as *Group 4: means the agent is probably not carcinogenic to humans.*

This category is used for agents or mixtures for which there is evidence, suggesting lack of carcinogenicity in humans and in experimental animals.

Caprolactam

Appendix

Website	Company/Organisation
www.acgih.org	The American Conference of Governmental Industrial Hygienists (ACGIH)
www.mindfully.org/Pesticide/ACSH-Koop.htm	ACSH American Council on Science and Health
www.aiha.org	AIHA The American Industrial Hygiene Association
www.AllPlasticBottles.org	All Plastic Bottles Collection Programs
http://www.AFPT.com	Alliance for Polyurethane Foam
www.Polyurethane.org	Alliance for the Polyurethanes Industry
www.apme.org	APME Association of Plastics Manufacturers in Europe
www.Plastics-car.com	Automotive Learning Center
www.bisphenol-a.org	Polycarbonate/BPAGlobalGroup
www.bpf.co.uk	BPF British Plastics Federation
www.ccohs.ca/oshanswers/chemicals/endocrine.html	CCOHS Canadian Centre for Occupational Health and Safety
www.nihs.go.jp/GINC/asia/japan/nihs//news/gincn	Chemicals Safety in Japan Committee on Sick House
http://cerhr.niehs.nih.gov	CERHR Centre for Evaluation of Risks to Human Reproduction

chem.sis.nlh.gov/chemidplus	ChemID Plus Advanced - website
www.hse.gov.uk/coshh/	Control of Substances Hazardous to Health (COSHH)
www.envirosense.org iaq@mindspring.com www.dehp-facts.com	ENVIROSENSE Consortium on IAQ
www.ecpi.org	ECPI European Council for Plasticizers and Intermediate
www.ecvm.org	ECVM European Council of Vinyl Manufacturers
http://www.eea.int http://glossary.eea.eu.int/EEAGlossary/	European Environmental Agency (EEA)
iaq@mindspring.com	ENVIROSENSE Consortium on IAQ
www.epa.gov	EPA Environmental Protection Agency
www.eupc.org	EuPC European Plastics Converters
www.cpsc.gov/businfo/fhsa.html www.cpsc.gov/BUSINFO/regsumfhsa.pdf	FHSA Federal Hazardous Substances Act
http://home.intekom.com/tm_info/ rw90113.htm>	Genetically Manipulated Food Web Site
www.hhinst.com www.healthybuilding.net www.indoorair.htm	HHI The Healthy House Institute
safety.science.tamu.edu/hmis.html	HMIS Hazardous Materials Information System
www.HandsonPlastics.com	Lessons for Teachers on Plastics
http://ntp-server.niehs.nih.gov	NTP National Toxicology Program - Deptartment of Health and Human Services
http://niehs.nih.gov	NIEHS National Institute of Environmental Health Services

www.cdc.gov/niosh	NIOSH National Institute for Occupational Safety and Health
http://www.npi.gov.au/database/ substance-info/profiles/6.html	The National Pollutant Inventory (NPI) of Australia
http://ntp-server.niehs.nih.gov	National Toxicology Program (NTP)
www.osha.gov	Occupational Safety and Health Administration (OSHA)
www.pagev.org.tr	Turkish Plastic Processors Association (PAGEV)
www.phthalates.org	Phthalate Information Centre
www.polyiso.org	Polyisocyanurate Insulation Association
www.Plasticresource.com	Plastics and the Environment
www.Plasticsinfo.org	Plastics and Health
www.Polystyrene.org	Polystyrene Packaging
http://www.plastics.org www.AmericanPlasticsCouncil.org	American Plastics Council (APC)
http://www.chemindustry.com/chemicals. html	PubChem website
http://www.scidev.net	The Science and Development Network
http://www.sprayfoam.org	Spray Polyurethane Foam Alliance
www.Vinylinfo.org	The Vinyl Institute
www.vinyl.org.au	The Vinyl Council
www.vinylbydesign.com	Vinyl in Design
http://www.chm.bris.ac.uk/safety/ Carcinogenetclist.htm	University of Bristol, UK - List of Substances which are Carcinogenic, Mutagenic or Toxic to Reproduction etc.
http://www.who.int/en/ http://www.aboutcancer.info/ Carcinogenics/carcinogenics.html	UN World Health Organisation

Glossary

ACCELERATOR: Chemical added to increase curing rate.

ACRYLICS: A generic name for polymers produced from acrylic acid/derivatives, mostly polymethylmethacrylate, or polyacrylonitrile fibre.

ACRYLONITRILE: Unsaturated hydrocarbon containing nitrogene (CH_2 CHCN), monomer of polyacrylonitrile; part of nitrile rubber.

ACTIVATOR: Chemical used to activate the curing reaction.

ADDITION POLYMERISATION: A method of polymerisation where monomers are linked together without the splitting off of water or any other simple molecules.

ADDITIVES: The material added (compounded) to a polymer during its processing to improve (or alter) certain properties, (without increasing the strength) (i.e., antistatic agents, stabilisers, colorants, plasticisers, etc).

AGEING: The effect on materials exposed to an environment for an interval of time.

ALKYDS: A special group of thermosetting resins prepared from polyhydric alcohols and polybasic acid or anhydrides - used mainly as lacquers.

ANTIOXIDANT: A 'sacrificial' chemical that prevents (or reduces) oxidative degradation of the resin and helps to maintain the properties.

AROMATIC POLYMER: A polymer with aromatic ring structure (i.e., polyesters).

ALLOY (POLYMERIC):	Two or more immiscible polymers together in a system, glued together by another component, to form a plastic resin with enhanced performance properties.
ANTI-BLOCKING AGENTS:	Additives used to prevent the adhesion of two touching layers of a film during its fabrication and/or storage.
ANTIFOGGING AGENT:	An additive used to prevent condensation of moisture (which resemble a fog) on glass and other transparent materials.
ANTIOXIDANTS and ANTIOZONANTS:	Sacrificial additives used to prevent any of the negative effects of oxygen and ozone on the resin materials.
ANTISTATS:	Additives that will eliminate or lessen static electricity.
A-STAGE:	The very early stage in the reaction of certain thermosetting resins where the molecular weight is low and the resin is still soluble in some liquids and system is still fusible.
BINDER:	A resin (or other material) used to hold particles together. It is the continuous phase in a reinforced plastic that provides mechanical strength.
BIOCIDES and FUNGICIDES:	Additives that are used to inhibit the growth of fungi, bacteria, marine organisms and similar living matter.
BIO POLYMERS (BIOLOGICAL POLYMERS):	Are inherently biodegradable special polymers that are produced by nature (plants, animals, micro-organisms), or synthetically from monomers that occur naturally.
BIOPLASTICS:	Are special plastics that contain at least one or more biopolymers.
BLENDING:	Mixing of two or more polymeric liquids (along with an appropriate compatibiliser) to prepare polymer blends or polymer alloys.
BLOCK COPOLYMER:	A copolymer where there are blocks of repeated sequences of polymeric segments of different chemical structure. The blocks can be either regularly alternating at random.

BLOWING and FOAMING AGENTS: Chemicals which upon addition to plastics (or rubbers) followed by heating, generate gases that result in the resin assuming a cellular structure.

B-STAGE: Used to describe an intermediate stage of reaction for a thermoset where the material softens when heated, and swells somewhat in the presence of certain liquids, but may not completely fuse or dissolve. Thermoset resins are usually supplied in this uncured state.

BUTADIENE: Unsaturated hydrocarbon monomer ($CH_2 CHCHCH_2$) of butadiene rubber.

CARBON BLACK: A black filler or pigment produced by the incomplete burning of natural gas or oil. It is widely used in the rubber industry (and in plastics to some extent) for their UV protection and outside weathering protection.

CARBON FIBRE: Fibres with about 94% carbon produced by pyrolysis of Rayon, polyacrylonitrile fibres or pitch.

CARBON NANOTUBES: Hollow tubular single or multi walled structures, with wall thickness of 0.07 nm and interlayer spacing of 0.34 nm.

CARCINOMA: In medicine, carcinoma is any cancer that arises from epithelial cells. It is malignant by definition: carcinomas invade surrounding tissues and organs, and may spread (metastasis).

CAS REGISTRY NUMBER: An internationally recognised unique numerical identification system for chemical compounds, polymers, biological sequences, mixtures and alloys.

CATALYST (INITIATOR):   A substance that can change the rate of a chemical reaction without changing its composition and without entering the final product structure.

CELLULAR PLASTIC: A plastic with a number of (connected or interconnected) stable gas cells (bubbles) that dispersed intentionally into its mass, having decreased densities (see also Foams).

CENTRAL NERVOUS SYSTEM (CNS):   Represents the largest part of the nervous system, and includes the brain and the spinal cord. It has a fundamental role in the control of behaviour.

CHLOROPRENE:	Monomer of chloroprene rubber (CR), containing chlorine.
COLLOIDAL:	A system where one substance is divided into minute particles (colloidal particles) and dispersed throughout a second substance. Particle diameters in the range of 1 to 1000 nm.
COLORANTS and PIGMENTS:	Additives that are used to change the color of the plastic. They can be powdery or a resin/color premix.
COMMODITY RESIN:	The high volume - low price resin group that are used commonly (i.e., PE, PP, PS, PVC).
COMPOSITE:	A solid structural substance produced by a combination of two or more materials that retain their identities. Typically, one of the materials combined is the strengthening agent, the other being the matrix (a thermoset or a thermoplastic resin). The word composite is also used for systems that are reinforced (reinforced: where cumulative properties are superior to the individual components) by addition of certain solid particles (i.e., short fibre composites - long fibre composites - continuous fibre composites).
COMPOUND:	A mixture of polymer and additives before processing.
COMPOUNDING:	Preparation of a compound.
CO-POLYMER:	A polymer resulting from polymerisation of two or more different monomers i.e., ter-polymer, tetra-polymer for two- and three monomers, respectively. (*see also*: **RANDON COPOLYMER, ALTERNATING COPOLYMER, GRAFT COPOLYMER, BLOCK COPOLYMER.**)
COUPLING AGENT:	A material that is used to form a chemical bridge between the resin and glass fibre or mineral filler. By acting as an interface, bonding is enhanced.
CROSSLINKING:	Chemical reaction between different polymeric molecules to form a covalently bonded strong three-dimensional systems. During crosslinking, the consistency of material passes from liquid to solid or from a softer solid to a much harder one. (*see also* **CURING.**)

CROSSLINKED POLYMER:	A polymer system where all polymer molecules are joined to each other by covalent bonds.
C-STAGE:	A term used to describe the final stage of a thermoset reaction where the material is now relatively insoluble and infusible.
CURE:	The completion of the crosslinking process during which a plastic or resin develops its full strength.
CURING:	Crosslinking (or vulcanising) a polymer (in general, applied to improve properties like modulus, -strength, -thermal stability and to decrease water absorption, etc.)
DEGRADATION:	In accordance with ASTM D883-00 '*Standard Terminology Relating to Plastics*', degradation is a deleterious change in the chemical structure, physical properties, or appearance of a plastic.
DEGRADABLE PLASTIC:	Is a plastic designed to undergo a significant change in its chemical structure under specific environmental conditions, resulting in a loss of some properties that may vary as measured by standard test methods appropriate to the plastic and the application in a period of time that determines its classification.
(Hydrolytically) DEGRADABLE PLASTIC:	Is a degradable plastic in which the degradation results from hydrolysis.
(Oxidatively) DEGRADABLE PLASTIC:	Is a degradable plastic in which the degradation results from oxidation.
(Photo) DEGRADABLE PLASTIC:	Is a degradable plastic in which the degradation results from action of natural light.
(Bio)-DEGRADABLE PLASTIC:	Is a degradable plastic in which the degradation results from the action of naturally occurring micro-organisms, such as bacteria, fungi and algae.
DETERIORATION:	In accordance with ISO, the International Standards Organisation, deterioration is a permanent change in the physical properties of a plastic shown by impairment of these properties.

DENSITY:	The weight per unit volume of a substance (usually in kg/m^3).
DIETHYLSTILBOESTROL (DES):	A drug, a synthetic oestrogen that was developed to supplement woman's natural oestrogen production. First routinely prescribed by physicians in 1938 and later for miscarriages or premature deliveries, originally considered effective and safe for both the pregnant woman and the developing baby. Later it is found that pregnant women given DES had just as many miscarriages and premature deliveries as the control group. In the USA, an estimated 10 million women were exposed to DES during 1938-1971.
DIELECTIC STRENGTH:	A measure of the voltage required to puncture a material, expressed in volts per mil of thickness.
DIFFUSION:	The movement of a gaseous/liquid/solid material in the body of a polymer (*see also* **PERMEABILITY**).
DIMENSIONAL STABILITY:	Ability of a substance or part to retain its shape subjected to varying degrees of temperature, moisture, pressure, or other stress.
DISPERSION:	A system with finely suspended particles of one material into the other.
DYES:	Intensely coloured, transparent, synthetic or natural chemicals that can dissolve in a resin.
ELASTOMER:	A generic name for polymeric substances with enhanced plastic-elastic behaviour, which is characteristic for vulcanised rubber-like synthetic or natural polymers. Elastomers, at room temperature, return rapidly to approximately their initial dimensions and shape even after substantial deformation by a weak stress and release of the stress.
ENGINEERING POLYMER:	A group of polymers with high tensile strengths (bigger than 40 MPa) exhibiting high stable performance at a continuous temperature above 100 °C, (i.e., PA, PBT/PET, PAr, POM/PC and PPE).

EPOXY:

A thermoset polymer system with one or more epoxide groups in it, used mainly as matrices in composites or as adhesives. Epoxides can be cured by amines, alcohols, phenols, carboxylic acids, etc.

ETHYLENE:

Unsaturated hydrocarbon monomer (CH_2-CH_2) for PE, ethylene-propylene rubber, EPM or EPDM. Also known as ethane.

ETHYLENE POLYMERS:

Polymers based on ethylene or copolymers of ethylene, where ethylene has the greatest share.

EXTENDER:

A material added to a plastic compound to reduce the amount of resin required per unit volume.

FIBRE:

A single homogeneous strand of material having a length of at least 5 mm, which can be spun into a yarn or roving. Also a thread like structure that has L >100 diameter with d = 100-130 μm.

FIBRE-GLASS:

Filaments from drawn molten glass (continuous, staple).

FIBRE-GLASS REINFORCEMENT (FGR):

Fibre-glass is the major economic material used for reinforcing - for both thermoset and thermoplastic - plastics (mat, roving, fabric).

FIBRE-GLASS REINFORCED (PLASTIC) (FRP):

The general term for a reinforced plastic (by using cloth-mat-stranded fibres).

FILAMENT:

Continuous polymeric structures (continuous fibres) with d = 100-130 μm.

FILMS:

Flat materials that are extremely thin (with a maximum nominal thickness of 0.25 mm), in comparison to their length and breadth.

FILLERS:

Relatively inert substances used with plastics and rubbers to improve physical, mechanical, thermal, electrical properties and for cost reduction purposes.

FIRE RETARDANTS:

Compounds which when mixed with a resin reduces the flammability.

FLAME RESISTANCE: Ability of a polymeric material to extinguish a flame after the source of heat is removed.

FLAME, FIRE and SMOKE RETARDANTS: Special chemicals used to reduce or inhibit the tendency of a polymer to burn, and decrease the smoke evolved.

FLAME RETARDED: A polymer system modified successfully by flame retardants so that it will not burn if exposed to a flame, or will self-extinguish if begins to burn.

FLAMMABILITY: A measure of a system for the extent of its support of combustion (ASTM D-635/UL E-94 tests).

FLUOROPOLYMERS: The family of fluorine-containing polymers, with good thermal and chemical resistances.

FOAM: Numerous cells produced by a blowing agent or by the reaction which are disposed throughout the mass (rigid, flexible, open, closed) (*see also* **CELLULAR**).

GEL: Crosslinked, swollen part of the polymer in liquid state. The irreversible point at which a polymer changes from a liquid to a semi-solid. Sometimes called the 'B' stage.

GELATION: The formation of a gel.

GLASS: A supercooled liquid, a hard and brittle material that solidifies from the molten state without crystallisation. Typical polymeric glasses are: atactic PS, PMMA, PC.

GLASS TRANSITION TEMPERATURE (T_g):   Characteristic temperature of the polymer where 20 to 50 main chain atoms move together in a cooperative way, the temperature where amorphous solid polymer changes from being vitrous to viscous; a temperature which is kinetic in nature, the value of which depends on the rate of heating or frequency of testing for the same polymer. For organic polymers, it is usually between −160 °C to 400 °C.

GRAFT COPOLYMER: A copolymer where chains consist of two or more parts of different composition, covalently joined to each other, one of which forms the main chain (polymer A), and the other(s) forms the side chains (polymer B).

GRAPHITE:	A crystalline allotrope of carbon.
GRAPHITE (FIBRE):	A substance (fibre) prepared from either pitch or polyacrylonitrile, precursors by oxidation-carbonisation and graphitisation processes.
GREEN CHEMISTRY:	Is the development of materials and processes that are environment-friendly.
HARDENERS:	Polyfunctional chemicals that can help to crosslink the system (in thermosets).
HEAT RESISTANCE:	The ability to resist the deteriorating effects of high temperatures.
HEAT STABILITY:	Resistance to any chemical deterioration caused by heat.
HEAT STABILISER:	A chemical that improves thermal resistance of the polymeric system.
HOMOPOLYMER:	A polymer with one type of repeating units, produced from a single type of monomer.
HYDROLYSIS:	Chemical decomposition of a chemical by the effect of water.
HYDROPHILIC:	Chemicals with strong dipole moments, being capable of absorbing water.
HYDROPHOBIC:	Chemicals capable of repelling water.
IMPACT MODIFIER:	Additives used to enhance the material's ability to withstand the force of impact.
INFRARED (IR):	Part between the visible light and radar range of the electromagnetic spectrum, which corresponds to radiant heat.
INHIBITOR:	A chemical which when used in small quantities can reduce or eliminate completely the systems reactivity and suppresses the chemical reaction.
INORGANIC:	A compound that is not composed of carbon atoms.

INORGANIC POLYMER:	A polymer composed of homopolar interlinkages between multi-valent elements other than carbon. Carbon-containing groups can be present in the side branches or as interlinks between structural members.
INTERFACE:	The boundary/surface between two different and physically distinguishable phases.
INTERPENETRATING POLYMER NETWORK (IPN):	A polymeric system with two or more different types of crosslinked polymers.
INTUMESCENCE:	The foaming and swelling of a plastic when exposed to high surface temperatures or flames.
IONIC POLYMERS (IONOMERS):	Polymers with linear or network structures containing ionic groups on the polymeric chains.
ISOTACTIC POLYMER:	A polymer where repeating units have the same stereochemical configuration.
KAUTSCHUK:	The German name for rubber arising from the Inka name 'cahu-chu' meaning 'the crying tree'.
LAMINATE:	A system composed of a series of layers (of polymer and reinforcement) bonded together.
LATEX, LATICES:	An aqueous dispersion of polymeric particles, a polymer emulsion.
LIGHT RESISTANCE:	The ability of a plastic material to resist fading, darkening or degradation after exposure to sunlight or ultraviolet light.
LIGHT, UV STABILISERS and ABSORBERS:	Additives that increase the ability of the material to withstand the negative effects of light and UV exposure, thus increasing the service life of the material.
LIQUID CRYSTAL POLYMER (LCP):	A thermoplastic polyamide or a polyester, with primary benzene rings in the backbone. They can be used at temperatures about 260 °C.
LUBRICANT:	**Internal lubricants** are chemicals that promote resin flow without affecting the fusion properties of a compound.

External lubricants are chemicals that promote release from metals which aid in the smooth flow of melt over die surfaces during processing.

MACROMER:	An oligomeric or telomeric chain that can polymerise.
MACROMOLECULE:	A polymer system with very large polymer chains, where end groups of chains or substitution of any group on the chain has no significant effect on material properties.
MASTERBATCH:	A concentrated blend of pigment, additives, fillers, and so on, in a base polymer.
MECHANICAL PROPERTIES:	Properties related to response of the material to stress/strain.
MELTING POINT:	Temperature when a resin changes from a solid to a liquid consistency.
MESOTHELIOMA:	Is an uncommon form of cancer, affecting the chest or abdomen, usually associated with previous exposure to asbestos.
MIL (MIL THICKNESS):	The unit used in measuring film thickness. One mil equals one thousandth of an inch (1 mil = 0.001 inch).
MISCIBLE POLYMER BLEND:	A polymer blend which is homogeneous down to the molecular level.
MOULD:	The tool used to fabricate the desired part shape. Also used to describe the process of making a part in a mould.
MOULDING:	The process of using a mould to form a part.
MOULD RELEASE:	A compound that is applied to the mould surface which acts as a barrier between the mould and the part, thus preventing the part from bonding to the mould.
MONOMER:	Small (gaseous or liquid) molecules that can attach to each other or to their similar reactive entities by covalent bonds (polymerisation reaction) to produce polymers. Derives from Greek word 'mono' meaning one and 'meros' meaning part.

NANOCOMPOSITE: Materials that contain nanometer-sized (10-9 m) particles in a matrix.

NATURAL RUBBER (NR): Rubber extracted from the tree, *Hevea brasiliensis.*

NETWORK POLYMER: Polymers obtained by the polymerisation of monomers having two or more functional groups which become interconnected with sufficient interchain bonds to fan a large 3-D network.

NITROSAMINES: Compounds that form by the reaction of nitrosating agents with amines, usually produced during vulcanisation of rubbers. Although there is no direct evidence for their carcinogenic effects, they are considered as 'suspected' agents.

NONPOLAR: Having no connection of electrical charges on a molecular scale, thus incapable of significant dielectric loss. Polystyrene and polyethylene are non polar.

NYLON: A trade name for polyamide.

POLYOLEFINS: Polymers produced from sulfur olefins, such as ethylene, propylene and so on.

OLIGOMER: Low molecular weight polymeric material, with only a few repeating units.

ORGANIC COMPOUND: A compound that is mainly composed of carbon atoms.

ORGANIC POLYMER: A polymer mainly composed of carbon atoms on the main chain.

OXIDATION: Chemical reaction with oxygen to produce a new compound (i.e., an oxide).

OXYGEN INDEX: The minimum volumetric percentage concentration of oxygen needed to just support combustion, at room temperature.

PERMEABILITY: Diffusion of a material from one side of the test piece towards the other side.

PHASE: Any separate portion of a system.

PHOTODEGRADATION:	Degradation caused by visible and ultraviolet, irradiation.
PIGMENT:	A dispersion (of undissolved particles) that impart colour to plastic.
PLASTIC:	A processed and shaped compound of a polymer and additives mixture.
PLASTICISER:	A compound of lower molecular weight that can reduce the T_g and increase the flexibility of a polymer.
PLASTISOL:	Mixtures of plasticisers and resins which can be converted to continuous films by applying heat. (Or a mixture of polyvinylchloride, with a plasticiser that can be moulded, and be cast.)
POLYMER:	A high molecular weight organic compound, natural or synthetic, whose structure can be represented by a repeated small unit, the mer, e.g., polyethylene, rubber, cellulose. The word 'polymer' derives from the Greek word 'poly' meaning many and 'meros' meaning part.
POLYMER ALLOY:	An immiscible polymer mixture with a modified interphase, interface and morphology.
POLYMER BLEND:	A mixture with two or more polymers or copolymers.
POLYMER:	A material composed of a number of monomers (mers) connected to each other with strong covalent bonds.
POLYMERIC NANOCOMOSITE:	A polymer/copolymer containing dispersed nanosized particles (10^{-9} m) in it.
POLYMERISATION:	A chemical reaction where monomer molecules join to each other to produce polymers.
POLYOLEFIN:	A polymer prepared by the polymerisation of an olefin(s) as the sole monomer(s).
PROPYLENE:	Unsaturated hydrocarbon monomer ($CH_2 CHCH_3$) of polypropylene, part of ethylene-propylene rubber.
PREPOLYMER:	A chemically active intermediate with sizes between the monomer and polymer.

RADICAL POLYMERISATION:	Polymerisation reaction where active centres are mainly radicalic.
RECYCLING:	Processing of used plastics and rubbers to either produce second quality materials, new chemicals and monomers, or energy.
REGRIND:	Reclaimed granulated waste material with or without the virgin material.
REINFORCEMENT:	A strong, inert woven or non-woven fibrous material used with polymers to increase the strength.
RELEASE AGENT:	A chemical added to a compound (or applied to the mould cavity, or both), to reduce parts sticking to the mould.
RIGID PLASTICS:	A plastic with elastic modulus greater than 690 MPa (in flexure or in tension), at 23 °C and 50% relative humidity.
RUBBER:	A generic term for elastomers and their compounds, or materials capable of quick recovery from large deformations.
SELF EXTINGUISHING:	A material that ceases to burn when the source of flame is removed.
SHELF LIFE:	The length of time a material can be stored under specified environmental conditions, continue to meet all applicable specification requirements and/or remain suitable for its intended function. The term is applied to finished products and raw materials.
SILICA:	A naturally occurring (white or colourless) mineral in, which is the basic component of quartz, sand or granite and so on.
SILICONES:	Chemicals derived from silica. They comprise alternating silica and oxygen atoms, with various organic groups attached to them.
SILICONE POLYMERS:	Silicon-based polymers (polyorganosiloxanes, such as: polydimethylsiloxane, silicone rubber).

SLIP AGENT:	A special additive used to provide surface lubrication during and immediately following processing of the plastic material. Slip agents act as an internal lubricant which will eventually migrate to the surface.
SMILES:	The Simplified Molecular Input Line Entry Specification. It is a specification used for describing the structure of chemical molecules using short ASCII strings.
SOLVENT:	A chemical, usually in liquid form, that dissolves another chemical.
STABILISER:	A compound that helps to maintain the physical and chemical properties of the polymeric system.
STABILISERS and SURFACE MODIFIERS:	Additives included in this category are 'antioxidants and antiozonants, antistats, biocides, and fungicides, heat stabilizers, light and UV stabilisers and absorbers'.
STRAIN:	The change in length per unit of original length during a deformation or flow process.
STRESS:	The ratio of applied load to the original cross section.
STYRENICS:	A group of polymers and copolymers produced from styrene.
STYRENE:	An aromatic hydrocarbon, monomer of polystyrene, part of styrene-butadiene rubber.
SUSTAINABLE DEVELOPMENT:	Development that provides for the needs of the present without compromising the ability of future generations to meet their own needs. (Definition of UN-World Commission on Environment and Development.)
TENSILE STRENGTH/ STRESS:	The maximum tensile load per unit area of the original cross section.
TERPOLYMER:	A copolymer composed of three different repeating units or monomers.
THERMOPLASTIC (TP):	Polymers that are capable of being repeatedly softened by heating and hardened by cooling.
THERMOPLASTIC ELASTOMER (TPE):	An elastomer with physical crosslinks that can turn into an uncrosslinked linear polymer upon heating.

THERMOSETS:	Infusible polymeric materials that can not soften or melt by heat, and can decompose at high temperatures without melting.
ULTRAVIOLET (UV):	The region of the electromagnetic spectrum between the violet end of visible light and X-ray range corresponding to between 10-390 nm, composed of high energy photons.
UNSATURATED POLYESTER (UP):	A low molecular weight (liquid) polyester with unsaturated double bonds, that can enter reaction with a monomer (i.e., styrene) to yield to a crosslinked thermoset system.
UV STABILISER:	A chemical that can selectively absorb UV rays and provides resistance to degradation from UV radiation by sacrificing itself.
VINYL CHLORIDE POLYMERS:	Polymers based on polyvinylchloride, or its copolymers.
VINYL (Vinylics):	Denotes mainly PVC and one or two other resins such as PVAc and PVDF.
VULCANISATION:	An irreversible process used to convert raw rubber compounds into a crosslinked elastomeric system, through sulfur or sulfur compounds. Arises from 'vulcan' being the God of the fire and the smithy.

Abbreviations

2,3,7,8-TCDD	Tetrachlorodibenzo-*p*-dioxins
AA	Acrylamide
ABS	Acrylonitrile butadiene styrene copolymer
ACGIH	American Conference of Governmental Industrial Hygienists
ACN	Acrylonitrile
ADC	Azodicarbonamide
AIHA	American Industrial Hygiene Association
ANSI	American National Standards Institute
APE	Alkylphenol ethoxylates
APME	Association of Plastics Manufacturers in Europe
ASA	Acrylonitrile styrene acrylate
ASTDR	Agency for Toxic Substances and Disease Registry
ASTM	American Society for Testing and Materials
ATBC	Acetyl triburyl citrate
ATH	Aluminium hydroxide
ATSDR	Agency for Toxic Substances and Disease Regulatory. US Dept. of Health and Human Services, Public Health Service
BBP	Butyl benzyl phthalate
BGA	Bundesgesundheitsamt
BHA	Butylated hydroxyl anisole
BHA-BHT	Butylated hydroxyl anisole – butylated hydroxytoluene
BHT	Butylated hydroxytoluene

BISGMA	Bisphenol A glycidyl dimethacrylate
BMI	Bis-maleimide
BNA	2-Naphthylamine
BOPP	Biaxially oriented PP
BPA	Bisphenol-A
BPADP	Bisphenol-A diphenyl phosphate
BPO	Benzoyl peroxide
BR	Butadiene rubber
BS	Butadiene styrene rubber
CA	Cellulose acetate
CAA	Clean Air Act
CAB	Cellulose acetate butyrate
CAP	Cellulose acetate propionate
CAS	Chemical Abstracts Service
CCA	Chromated copper arsenate
CCOHS	Canadian Centre for Occupation Health and Safety
CD	Compact disk(s)
CDC	Centres for Disease Control and Prevention
CD-ROM	Compact disk - read only memory(s)
CERF	Constructing an Efficient and Renewable Future
CFC	Chlorofluorocarbons
CFR	US Code of Federal Regulations
CFRP	Carbon fibre reinforced plastic
CIIT	Chemical Industry Institute of Toxicology
CIS	Carcinoma *in situ*
CN	Cellulose nitrate
CNS	Central nervous system
COBRAE	Composites Bridge Alliance Europe - The Netherlands
COPE	Copolyester

COSHH	Control of Substances Hazardous to Health
CP	Cellulose propionate
CPE	Chlorinated PE
CPVC	Chlorinated PVC
CR	Chloroprene Rubber
CSIRO	Commonwealth Scientific and Industrial Research Organisation
CWA	The Clean Water Act
DBP	Dibutyl phthalate
DCDD	Dichlorinated dioxins
DCDMH	1,3-Dichloro-5,5-dimethylhydantoin
DDS	4,4′-Sulfonyldianiline
DEHA	Di-2-ethylhexyl adipate
DEHP	Di-2-ethyl hexyl phthalate
DES	Diethylstilboestrol
DETA	Diethylenetriamine
DGE	Diglycidyl ether
DGEBA	Diglycidyl ether of bisphenol A
DHEW	US Department of Health, Education and Welfare
DHI	Dihydroindolizine
DHP	Di-*n*-hexyl phthalate
DIBP	Di-isobutyl phthalate
DIDP	Di-isodecyl phthalate
DIHP	Di-isoheptyl phthalate
DINP	Di-isononyl phthalate
DMEP	Dimethoxyethyl phthalate
DMFA	Dimethylformamide
DMHA	Dimethylhexyl adipate
DMHP	Dimethyl hydrogen phosphite
DMMP	Dimethyl methylphosphonate

DMPh	Dimethyl phthalate
DMSO	Dimethyl sulfoxide
DMT	Dimethyleterephthalate
DNA	Deoxyribonucleic acid
DNOC	Dinitro-*o*-cresol
DNOP	Di-*n*-octylphthalate
DOA	Dioctyl adipate
DOP	Dioctyl phthalate
DPA	Diphenylamine
DPGME	Dipropylene glycol methyl ether
DPK	Dipropyl ketone
DVB	Divinyl benzene
DVD	Digital versatile disk(s)
EC	Ethyl cellulose
ECD	Endocrine disrupter chemical(s)
ECH	Epichlorohydrin
ECPI	The European Council for Plasticisers and Intermediates - Belgium
EDC	Epigenetic
EDC	Epigenetic carcinogen
EEA	European environmental agency
EIPRO	Environmental Impact of Products
EP	Epoxide or epoxy
EPA	US Environmental Protection Agency
EPDM	Ethylene-propylene diene terpolymer
EPE	Expanded polyethylene
EPM	Rubbers of ethylene-propylene monomer
EPR	Ethylene-propylene rubber
EPS	Expanded polystyrene
ESD	Electrostatic-discharge dissipating

ETU	Ethylene thiourea
EU	European Union
EU-COM	Community Strategy for Endocrine Disruptors
EuP	Energy-using products
EVAc	Ethylene-vinyl acetate
EVOH	Ethylene-vinyl alcohol
FDA	Food and Drug Administration
FEP	Fluorinated ethylene propylene
GC/DMS	Gas chromatography/differential mobility spectroscopy
GC/MS	Gas chromatography/mass spectroscopy
GFRP	Glass fibre reinforced plastic
HALS	Hindered amine light stabilisers
HCFC	Hydrochlorofluorocarbons
HCN	Hydrogen cyanide
HCRA	Harvard Centre for Risk Analysis
HDI	Hexamethylene diisocyanate
HDPE	High-density polyethylene
HEMA	Hydroxy ethyl methacrylate
HIPS	High impact polystyrene
HITEC	The Highway Innovative Technology Evaluation Center
HMDI	Hydrogenated MDI
HMPA	Hexamethyl phosphoramide
HNBR	Hydrogenated nitrile rubber
HPA	2-Hydroxypropyl acrylate
HRV	Heat recovery ventilator(s)
HVAC	Heating-ventilation and air conditioning system
IAOMT	The International Academy of Oral Medicine and Toxicology
IAQ	Indoor Air Quality
IARC	International Agency for Research on Cancer

IGE	Isopropyl glycidyl ether
IMPRO	Environmental Improvement of Products
IPDI	Isophorone diisocyanate
IPP	Integrated performance primitives
IR	Infrared radiation
ISO	International Organisation for Standardization
IV	Intravenous
LCP	Liquid crystal polymer(s)
LD	Lethal concentration
LD50 or MLD	Median lethal dose
LDPE	Low-density polyethylene
LLDPE	Linear low-density polyethylene
MA	Maleic anhydride
MAA	Methacrylic acid
MAc	Methyl acrylate
MAC	Methyl acrylate
MAN	Methacrylonitrile
MAP	Modified atmosphere packaging
MBK	Methyl butyl ketone
MBT	2-Mercaptobenzothiazole
MDA	4,4′-Methylene dianiline
MDF	Medium density fibreboard
MDH	Magnesium hydroxide
MDI	Methylene diisocyanate
MDPE	Medium density PE
MEHP	Monoethylhexyl phthalate
MEK	Methylethyl ketone
MF	Melamine-formaldehyde
MIAK	Methyl isoamyl ketone

MIBK	Methyl isobutyl ketone
MIC	Methyl isocyanate
MLD or LD 50	Median Lethal Dose
MMA	Methyl methacrylate monomer
MNBK	Methyl butyl ketone
MTHPA	Methyl tetrahydro phthalic anhydride
MW	Molecular weight
NASA	National Aeronautics and Space Administration
NBR	Nitrile rubber
NBS	National Bureau of Standards
NDI	Naphthalene diisocyanate
NDPhA	*N*-Nitrosodiphenylamine
NGCC	Network Group for Composites in Construction - UK
NIH	National Institute of Health
NIH-DHHS	National Institute of Environmental Health Sciences
NIOSH	National Institute for Occupational Safety and Health
NIPA	*N*-Isopropylaniline
NMOR	*N*-Nitrosomorpholine
NR	Natural rubber
NRL	Natural rubber latex
NTP	National Toxicology Program
OBPA	10-10´-Oxybisphenoxarsine
OBSH	4,4-Oxybis benzene sulfonyl hydrazide
OECD	Organisation for Economic Co-operation and Development
OITO	2-*N*-Octyl-4-isothiazolin-3-one
ORD	Office of Research and Development
OSHA	Occupational Safety and Health Administration
P(nBMA)	Poly-*n*-butylmethacrylate
PA	Polyamide(s)

PAA	Polyacrylamide
PAEK	Polyaryletherketone
PAI	Polyamide-imide
PAN	Polyacrylonitrile
PB	Polybutene-1
PBA	Polybutylene acrylate
PBB	Polybrominated biphenyls(s)
PBD	Polybutadiene
PBDE	Polybrominated diphenyl ether
PBNA	N-Phenyl-β-naphthylamine
PBT	Polybutylene terephthalate
PBT-S	Saturated polybutylene terephthalate
PBT-U	Unsaturated PBT
PC	Polycarbonate(s)
PCB	Polychlorinated biphenyl(s)
PCDD	Polychlorinated dibenzo-p-dioxins
PCDF	Dibenzofuran(s)
PCM	Perchloromethyl mercaptan
PCTEE	Polychlorotrifluoroethylene
PE	Polyethylene
PEEK	Polyetheretherketone
PEG	Polyethylene glycol
PEI	Polyether-imide
PEK	Polyetherketone
PEN	Polyethylene naphthanate
PES	Polyethyl sulfone
PET	Polyethylene terephthalate
PET-S	Saturated polyethylene terephthalate
PET-U	Unsaturated PET

PF	Phenol-formaldehyde
PFOA	Perfluoro-octanoic acid
PGME	Propylene glycol monomethyl ether
phr	Parts per hundred rubber (or plastic)
PI	Polyimide
PIB	Polyisobutylene
PLA	Polylactic acid
PMMA	Polymethylmethacrylate
PMP	Poly-4-methylpentene-1
PO	Propylene oxide
POM	Polyoxymethylene
POM	Polyoxymethylene
POP	Persistent organic pollutants
PP	Polypropylene(s)
PPA	Polymer processing additive(s)
ppb	Parts per billion
PPD	*p*-Phenylene diamines
PPE	Polyphenylene ether
ppm	Parts per million
PPO	Polyphenylene oxide
PP-PS	Polypropylene-polystyrene
PPS	Polyphenylene sulfide
PPSU	Polyphenylene sulfone
PS	Polystyrene
PS-BD	Polystyrene-polybutadiene
PSU	Polysulfone
PTFE	Polytetrafluoroethylene
PU	Polyurethane(s)
PVA	Polyvinyl alcohol

PVAc	Polyvinyl acetate
PVB	Polyvinyl butyral
PVC	Polyvinyl chloride
PVC-P	Plasticised PVC
PVDC	Polyvinylidene chloride
PVDF	Polyvinylidene fluoride
PVFM	Polyvinyl formal
PVK	Polyvinylcarbazole
RCRA	Resource Conservation and Recovery Act
RDP	Resorcinol diphenyl phosphate
REACH	Registration Evaluation and Authorisation of Chemicals
RNA	Ribonucleic acid
RoHS	Restriction of the Use of Certain Hazardous Substance Directive
RoHS	Restriction of the Use of Certain Hazardous Substances
SAMPE	Society for the Advancement of Material and Process Engineering
SAN	Styrene acrylonitrile copolymer
SARA	Superfund Amendment and Reauthorisation Act
SB	Styrene-butadiene
SBR	Styrene butadiene rubber
SBS	Styrene-butadiene-styrene
SCF	Scientific Committee on Food
SCHER	EU Scientific Committee on Health and Environmental Risks
SD	Sustainable Development
SDS	Sustainable development strategy
SIS	Styrene-isoprene-styrene
SML	Specific migration limit(s)
SPE	Society of Plastics Engineers, Inc
SPI	Society of the Plastics Industry, Inc
STEL	Short-term exposure limit

SVOC	Semi-Volatile Organic Compounds
TBBC	4,4'-Thiobis(6-*tert*-butyl-*m*-cresol)
TBP	Tributyl phosphate
TBPBA	Tetrabromophenol *bis* phenol A
TCDD	Tetrachlorinated dioxins
TCR	Tricrescyl
TDI	Toluene diisocyanate
TEF	Toxicity Equivalent Factor
TEGDMA	Triethylene glycol dimethacrylate
TEHTM	Tri-(2-ethylhexyl)trimellitate
TET	Triethyl-tin
TETA	Triethylene tetramine
TGA	Thioglycolic acid
TGMDA	Tetraglycidyl-4-4'-methylene dianiline
TGPAP	Triglycidyl-*p*-amino-phenol
THF	Tetrahydrofuran
TLV	Threshold Limit Value
TLV-C	Threshold limit value - Ceiling
TLV-STEL	Threshold limit value - short term exposure limit
TLV-TWA	Threshold limit value-time weighed average
TMA	Trimellitic anhydride
TMB	Trimethyl benzene
TMP	Trimethyl phosphite
TMSN	Tetramethyl succinonitrile
TMT	Trimethyl tin
TMTD	Thiram
TOCP	Triorthocresyl phosphate
TOF	Trioctyl phosphate
TPA	Terephthalic acid

TPE	Thermoplastic elastomer(s)
TPO	Thermoplastic polyolefin
TPP	Triphenyl phosphate
TPP	Triphenyl phosphite
TPPE	Thermoplastic polyester
TPU	Thermoplastic polyurethane(s)
TrCDD	Trichlorinated dioxins
TSH	*p*-Toluene sulfonyl hydrazide
TVOC	Total VOC
TWA	Time weighted average
UF	Urea-formaldehyde
UHMWPE	Ultra-high molecular-weight polyethylene
UN	United Nations
UN/ECE	United Nations Economic Commission for Europe
UP	Unsaturated polyester(s)
uPVC	Unplasticised PVC
UPVC	Unplasticised PVC
US-FDA	United States Food and Drug Administration
USP	United States Pharmacopoeia
UV	Ultraviolet
VCD	Vinyl cyclohexene dioxide
VCH	4-Vinylcyclohexene
VCM	Vinyl chloride
VDC	Vinylidene chloride
VOC	Volatile organic compound(s)/chemical(s)
VVOC	Very Volatile Organic Compounds
WEEE	Waste Electrical and Electronic Equipment
WHO	World Health Organization
WPC	Wood-plastic composites

XLPE	Crosslinked PE
XLPE	Thermoset crosslinked polyethylene
ZDC	Zinc dithiocarbamates

Index

Titles of Related Interest

Smithers Rapra also publishes the following titles, which you may be interested in:

Health and Safety in the Rubber Industry
N. Chaiear, 2001
Rapra Review Report Number: 138
ISBN: 978-1-85957-301-3

Medical Polymers 2006
Smithers Rapra Conference Proceedings, 2006
ISBN: 978-1-85957-580-2

Pharmaceutical Applications of Polymers for Drug Delivery
David Jones, 2004
Rapra Review Report Number: 174
ISBN: 978-1-85957-479-9

Pharmaceutical Polymers 2007
Smithers Rapra Conference Proceedings, 2007
ISBN: 978-1-84735-017-6

Polymers in Medical Applications
B.J. Lambert, F-W. Tang and W.J. Rogers, 2001
Rapra Review Report Number: 127
ISBN: 978-1-85957-259-7

Practical Guide to Polyvinyl Chloride
Stuart Patrick, 2005
ISBN: 978-1-85957-511-6

The Role of Poly(Vinyl Chloride) in Healthcare
Colin Blass, 2001
ISBN: 978-1-85957-258-0

Sterilisation of Polymer Healthcare Products
Wayne Rogers, 2005
ISBN: 978-1-85957-490-4

Toxicology of Solvents
M. McParland and N. Bates, 2002
ISBN: 978-1-85957-296-2

To purchase any of these titles, or to view our extensive catalogue, please visit:
www.polymer-books.com

Orders or queries may also be directed to our publication sales team.
Email: publications@rapra.net *Tel*: +44 (0) 1939 250383 *Fax*: +44 (0) 1939 251118

Plastics, Rubber and Health

Printed in the United States
140282LV00007B/5/A